T0100431

# MATLAB for Engineering Applications

# MATLAB for Engineering Applications

Editor: Natalie Coffman

MURPHY & MOORE
www.murphy-moorepublishing.com

www.murphy-moorepublishing.com

# ⓂMURPHY & MOORE

**Cataloging-in-publication Data**

MATLAB for engineering applications / edited by Natalie Coffman.
   p. cm.
Includes bibliographical references and index.
ISBN 978-1-63987-694-5
1. MATLAB. 2. Engineering--Data processing. 3. Numerical analysis--Data processing.
I. Coffman, Natalie.
TA345.5.M42 M38 2023
510.285 536--dc23

© Murphy & Moore Publishing, 2023

Murphy & Moore Publishing
1 Rockefeller Plaza,
New York City,
NY 10020, USA

ISBN 978-1-63987-694-5

This book contains information obtained from authentic and highly regarded sources. Copyright for all individual chapters remain with the respective authors as indicated. All chapters are published with permission under the Creative Commons Attribution License or equivalent. A wide variety of references are listed. Permission and sources are indicated; for detailed attributions, please refer to the permissions page and list of contributors. Reasonable efforts have been made to publish reliable data and information, but the authors, editors and publisher cannot assume any responsibility for the validity of all materials or the consequences of their use.

**Trademark Notice:** Registered trademark of products or corporate names are used only for explanation and identification without intent to infringe.

# Contents

# Preface

MATLAB refers to a multi-paradigm programming language and numeric computing environment. It permits implementation of algorithms, plotting of functions and data, matrix manipulations, formation of user interfaces, and interfacing with programs written in other languages. It is a tool which allows the user to program, compute, and visualize the results graphically. It is mainly designed for numerical computation. It is an optional toolbox that utilizes the MuPAD symbolic engine that provides access to symbolic computing capabilities. MATLAB is widely used within various fields such as economics, engineering, and science. It can be used to simulate diverse electrical networks. Modern developments in the MATLAB has also made it a very competitive tool for wireless communication, data analytics, image processing, artificial intelligence, machine learning, and robotics. This book contains some path-breaking studies outlining the engineering applications of MATLAB. It will serve as a reference to a broad spectrum of readers.

This book unites the global concepts and researches in an organized manner for a comprehensive understanding of the subject. It is a ripe text for all researchers, students, scientists or anyone else who is interested in acquiring a better knowledge of this dynamic field.

I extend my sincere thanks to the contributors for such eloquent research chapters. Finally, I thank my family for being a source of support and help.

**Editor**

# MATLAB Co-Simulation Tools for Power Supply Systems Design

Valeria Boscaino and Giuseppe Capponi
*University of Palermo*
*Italy*

## 1. Introduction

Modern electronic devices could be thought as the collection of elementary subsystems, each one requiring a regulated supply voltage in order to perform the desired function. Actually, in a system board you can find as many as ten separate voltage rails, each one rated for a maximum current and current slew-rate. If the available energy source is not suitable for the specific application, a power supply system is required to convert energy from the available input form to the desired output. Since power-up, power-down sequences and recovery from multiple fault conditions are usually offered, monitoring and sequencing the voltage rails is quite complex and a central power management controller is used. The design of the power management system is a key step for a successful conclusion of the overall design. The cost, size and volume of the electronic device are heavily affected by the power supply performances. The proliferation of switching converters throughout the electronic industry is a matter of fact. Unlike linear regulators, switching converters are suitable for both reducing and increasing the unregulated input voltage and efficiency can be as high as 97%. Classification of switching power supplies is mainly based on the available input energy form and the desired output: AC-AC cycle conversion: alternate input, alternate output; AC-DC rectification: alternate input, continuous output; DC-AC inversion: continuous input, alternate output; DC-DC conversion: continuous input, continuous output. Several topologies are well suited to perform the required conversion (Erikson et al., 2001; Pressman et al., 2009). Independently of the specific topology, the architecture of a power supply system can be generalized as shown in Fig.1. A power section and a control section are highlighted. The switching converter is included in the power section while the feedback network and protection circuits for safe operation are collected in the control system. Mainly because of the growing interest in digital controllers, a power supply system can be modelled as a complex system, quite often involving both digital and analog subsystems. If compared to their analog counterparts, digital controllers offer lower power consumption, higher immunity to components aging and higher and higher design flexibility. The control algorithm is often described at a functional level using hardware description language, as VHDL or Verilog. ASICs or FPGA based controllers are usually integrated by the means of sophisticated simulation, synthesis and verification tools. Time to market is heavily reduced by the means of powerful simulation environments and accurate system modelling. Achieving high accuracy models for mixed-mode systems is quite difficult. Specific simulation tools are available on the market, each one oriented to a specific abstraction level. Circuit simulation software as *Powersim PSIM* and *Orcad Pspice* are the most common choice for circuit modelling. In (Basso, 2008), the design and simulation of switch-mode power supplies is deeply analyzed and simulation tips in several environments are proposed. ASIC simulation and verification tools as Xilinx ISE/Modelsim or Aldec Active-HDL are available to implement the digital controller by the VHDL or VERILOG source code. Since the interaction between subsystems is the most common source of faults, testing separately analog and digital subsystems by the means of different verification tools is a severe mistake. Matlab is a powerful simulation environment for mixed-mode systems modelling and simulation, providing several tools for system co-simulation (Pop, 2010; Kaichun et al., 2010). The Matlab suite offers co-simulation tools for *PSIM*, *Modelsim* and *Active-HDL* simulation environments. Ideally, modelling the power converter by a circuit implementation in PSIM environment and the digital controller by the VHDL code in *Xilinx ISE* or *Active-HDL* environment allows the designer to test the composite system in Matlab environment using co-simulation procedures. Unfortunately, the use of several co-simulation tools in the same Simulink model heavily reduces the processing speed. As an example, in this chapter the Simulink model of a multiphase dc-dc converter for VRMs applications is described. The control system is described by VHDL language and the controller model is implemented in *Active-HDL* environment. For the highest processing speed, an alternative modelling technique for the power section is proposed by the means of elementary library blocks, avoiding *PSIM* co-simulation and not affecting the accuracy of behavioural simulations. As shown by simulation results, the high accuracy relies in the opportunity to match the system behaviour both within the switching event and during long-time events such as load transients and start-up. The great potential of the co-simulation procedure for mixed-mode systems is highlighted by the comparison between simulation and experimental results.

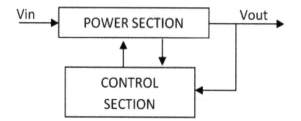

Fig. 1. Power supply system architecture.

## 2. Multiphase dc-dc converters for VRM applications

The evolution in microprocessor technology poses new challenges for power supply design. The end of 2009 marked the birth of 32nm technology for semiconductor devices, and 22nm is expected to be reached in the 2011-2012 timeframe. The next generation of computer microprocessors will operate at significantly lower voltages and higher currents than today's generation in order to decrease the power consumption and increase the processing speed. Within several years, Intel cores are expected to operate on a 0.8V supply voltage. High-quality power is delivered to the microprocessor by a point-of-load (POL) converter, also known as voltage regulator module (VRM), which is located on the motherboard next to the load. Embedded POL converters are designed to supply a tight regulated output voltage in the 0.8 – 1.2V range. Due to integration advances and increasing clock frequencies, high current values, often higher than 100A, are required. VRMs supply highly dynamic loads exhibiting current slew rates as high as 1000A/µs. Today, in distributed power architectures, individual POL converters usually draw power from a 12V intermediate power bus. Lowering the core voltage results in a very narrow duty cycle of the 12V-input converter, worsening the converter efficiency and load transient response. The best topology for VRM applications features tight voltage regulation, very fast transient response, high-current handling capabilities and high efficiency. Modern technology advances constantly prompt electronics researchers to meet all requirements investigating innovative solutions. The evolution from a classical buck topology to multiphase architectures is described to better understand the great advantages brought by multiphase architectures.

The buck topology and the MOSFET gate drive signal are shown in Fig.2. The gate signal period is always referred to as the converter switching period T. The duty-cycle D is defined as the ratio between the MOSFET on-time and the switching period. Since one terminal of the power MOSFET is tied to the high-side of the input voltage rail, the MOSFET is also called high-side MOSFET. The steady-state analysis is here performed assuming constant the input voltage and the output voltage during a switching period.

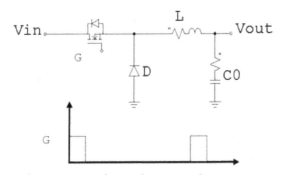

Fig. 2. The buck converter topology and gate drive signal.

On the rising edge of the gate drive signal, the high-side MOSFET is turned on. The inductor is now series connected between the input and output rails and the diode is reverse biased. During the MOSFET on-time, the inductor current slope is constant and it can be expressed as:

$$S_{on} = \frac{Vin - Vout}{L} \tag{1}$$

On the falling edge of the gate drive signal, the MOSFET is turned off. The inductor opposes the current break trying to maintain the previously established current. The voltage polarity across it immediately reverses. Since the right terminal of the inductor is tied to the output rail, the cathode voltage of the diode D is driven toward negative values. When zero crossing occurs, the diode D is forced into conduction clamping the cathode node to ground. The inductor voltage is now reversed and full inductor current flows through the diode D. The inductor is now connected in parallel to the output voltage rail. During the OFF-time, the slope of the inductor current is given by:

$$S_{off} = \frac{-Vout}{L} \tag{2}$$

Under steady-state conditions, the average inductor voltage over a switching period is zero, resulting in the expression:

$$(Vin - Vout) \cdot T_{on} = -Vout \cdot T_{off} \tag{3}$$

Or

$$Vout = Vin \cdot \frac{T_{on}}{T_{on} + T_{off}} \tag{4}$$

where Ton is the MOSFET conduction time and Toff the MOSFET off-time.
As shown in Fig.3 the inductor undergoes a magnetizing cycle over the entire switching period. The inductor current ripple could be expressed as:

$$\Delta I_L = \frac{Vin - Vout}{L} \cdot T_{on} = \frac{-Vout}{L} \cdot T_{off} = \frac{Vin(1-D)D}{L} T \tag{5}$$

where D is the duty cycle and T the switching period.

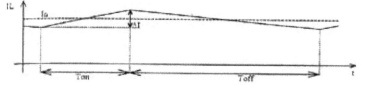

Fig. 3. The inductor current waveform over the switching period under steady-state conditions.

Since at steady-state the average capacitor current is null, the average inductor current is equal to the load current. The ac component of the inductor current flows through the parallel connection between load and capacitor. Commonly, the capacitor is large enough to assume that its impedance at the switching frequency is much smaller than the load resistance. Hence nearly all the inductor current ripple flows through the capacitor. The ac current flowing into the load could be reasonably neglected. A continuous conduction mode is assumed. The inductor is sufficiently large to assume:

$$\frac{\Delta IL}{2} \ll I_{out} \tag{6}$$

The inductor current never crosses zero during the switching period. A continuous conduction mode is preferred in high-current applications since lower and lower conduction power losses are involved.
The capacitor current waveform is shown in Fig.4. When the capacitor current is positive, the stored charge as well as the capacitor voltage will increase. Vice versa, when the capacitor current goes negative, stored charge will be released and the output voltage will decrease as well.
The total voltage change occurs between two consecutive zero-crossings of the capacitor current. Evaluating the total charge q as the grey-coloured area in Fig.4 yields:

$$q = \frac{1}{2} \frac{\Delta I_L}{2} \frac{T}{2} \tag{7}$$

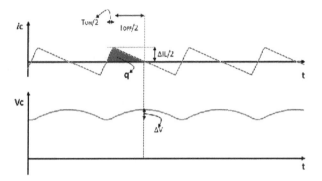

Fig. 4. At the top, the output capacitor current. At the bottom, the output voltage.

The voltage ripple across the output capacitor is given by:

$$\Delta V = \frac{q}{C} = \frac{\Delta I_L T}{8C} \tag{8}$$

Increasing the output capacitor value or the switching frequency directly improves the output voltage regulation. If a larger inductor is selected, the ripple current would decrease as well as the output voltage ripple. Yet, by increasing the inductor value the load transient response is worsened. A load transient could be modelled by the equivalent circuit shown in Fig.5.

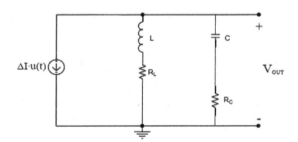

Fig. 5. The equivalent circuit under large-signal load transients.

The load transient is modelled as an ideal current step. Since the input voltage is assumed constant, the switching node is dynamically tied to ground. The inductor is thus parallel connected to the output capacitor filter. If the loop response is sluggish, the peak output voltage deviation $\Delta V_{out,max}$ tends to approach the open-loop value which is given by:

$$\Delta V_{out,max} = \Delta I \cdot \sqrt{\frac{L}{C}} + Rc \cdot \Delta I \tag{9}$$

where $\Delta I$ is the current step width and $Rc$ the equivalent series resistor of the output capacitor.

In Fig.6 the load current $Iload$, the inductor current $IL$, the capacitor current $Ic$ and the output voltage $Vout$ under a load transient are shown. By increasing the output capacitor value, the dynamic response could be further improved. Yet, in high-current applications a too large capacitance will be obtained affecting the system cost and size.

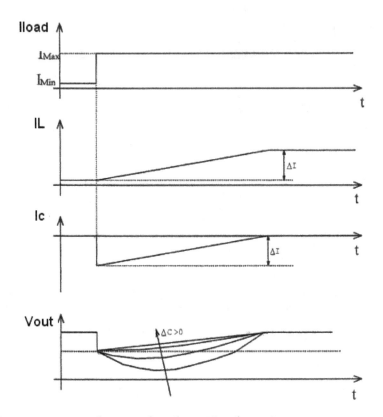

Fig. 6. The converter waveforms under a large-signal transient.

The optimization of steady-state and dynamic performances leads to a trade-off in the choice of the inductor value. As will be further discussed, the inductor value is usually chosen to meet steady-state requirements and dynamic performances are improved by the control algorithm.

As shown by (4), the buck converter reduces the dc input voltage by the means of switching elements. Power loss due to switches is ideally zero. Yet, switches are not ideal. Further, switching losses cannot be neglected. The inductor current flows through the MOSFET during the on-time and through the diode during the OFF time. Conduction losses are proportional to the forward voltage drop across the active elements. The MOSFET forward voltage drop is lower than the diode threshold voltage. In a buck converter, conduction losses are mainly due to the freewheeling diode.

In high-current low-voltage applications, synchronous rectification is widely used to improve efficiency. As shown in Fig.7, the diode D is replaced by a MOSFET featuring a very low forward voltage drop. The rectification function, typically performed by diodes, is now achieved by the means of a MOSFET which is driven synchronously with the high-side MOSFET. For this reason, the added MOSFET is sometimes referred to as the synchronous rectifier MOSFET. Since the rectifier MOSFET source terminal is tied to the low-side of the input voltage rail, the MOSFET is also called low-side MOSFET. If compared with classical topology, synchronous rectification heavily reduces conduction losses improving efficiency. During the conduction time of the high-side (on time) the low-side is off and vice versa.

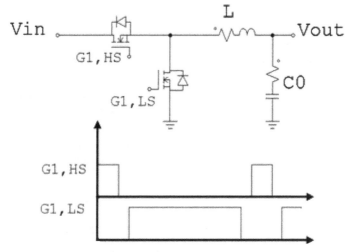

Fig. 7. The synchronous buck topology and MOSFETs gate drive signals.

A dead-time is required to avoid the simultaneous conduction of high-side and low-side MOSFETs. During the dead time, neither the low-side nor the high side MOSFET is on. When the high-side MOSFET is turned off, the inductor opposes the current break forcing the low-side MOSFET body-diode into conduction. Over the entire dead time the full inductor current flows through the body diode itself. At the end of the dead time, the low-side MOSFET is turned on, the body diode is bypassed by the active MOSFET channel and the inductor current flows through the low-side MOSFET channel. During the dead time, classical and synchronous topologies are exactly alike. It would be advisable a further reduction of both the diode threshold voltage and the dead time in order to limit power losses. A Schottky diode is often integrated in parallel to power MOSFET: the body diode will be bypassed by the Schottky diode over the entire dead time lowering the effective threshold voltage. A minimization of the low-side body-diode conduction time is required to take advantage from synchronous rectification. Sophisticated dead-time control algorithms are usually implemented by synchronous MOSFET drivers. If a single buck converter topology were employed in high-current applications, a large amount of output-filter and on-board decoupling capacitance would be required to achieve the expected transient response. The size and cost of the voltage regulator would increase as well. Multi-phase converters are widely used. Ripple cancellation and improved transient response are only a few advantages of multi-phase architectures. A detailed analysis of multiphase converters is reported in (Miftakhutdinov, 2001). In N-modules architecture, N identical buck converters are connected in parallel to the load resulting in a heavy reduction of the output ripple. Smaller inductances are allowed thus improving the transient response. Note that increasing the number of modules emphasizes the advantages of a multiphase architecture. Gate drive signals are interleaved to take full advantage from the multiphase architecture. Assume that each channel operates at a switching frequency $f_{sw}$. Under steady-state conditions, the phase shift between gate drive signals of each module depend on the number of interleaved channels. If an N-channels architecture is considered, assume that the control signal sequence is Phase 1, 2,.., N. The interleaved technique requires a 360°/N phase shift between two following channels. The maximum number of channel is limited by the relationship:

$$(1 - D) > N \cdot D \tag{10}$$

where N is the number of channel and D is the duty-cycle. A key feature of multiphase architecture is now highlighted: the ripple frequency of the output voltage is N-times the switching frequency of each power module. The equivalent switching frequency is higher and higher than the effective switching frequency of each module. The designed multiphase converter features a 4-modules architecture, as shown in Fig.8. Inductors and capacitors are modelled including the parasitic equivalent series resistance. Assume that the control sequence is Phase 1,2,3,4. Interleaved gate drive signals are shown in Fig.9. Fig.10 shows the inductor current waveforms for each buck channel. As shown in Fig.11, the equivalent current is defined as:

$$IL_{eq} = \sum_{i=1}^{n} IL_i = IL_1 + IL_2 + IL_3 + IL_4 \tag{11}$$

Ripple cancellation derives from the interleaving technique. The steady-state analysis is carried on within a fundamental Tsw/4 time interval. Within a Tsw/4 period, only one module could be driven in the ON state, the others are constantly OFF. For example, from 0 to Tsw/4 modules 2, 3 and 4 are constantly OFF while the first module (phase 1) is driven in the ON state during the subinterval [0, DTsw/4]. No matter which phase is active in the ON state, the fundamental period of the output waveforms is Tsw/4 since we have assumed identical modules.

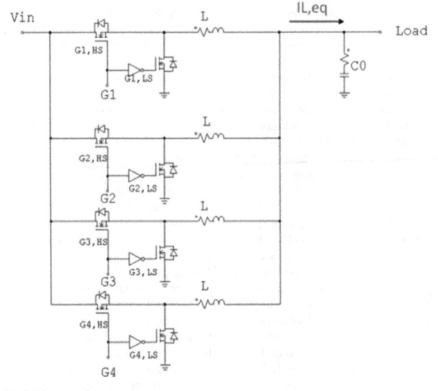

Fig. 8. Multiphase architecture.

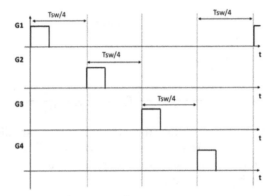

Fig. 9. High-side MOSFETs gate drive signals.

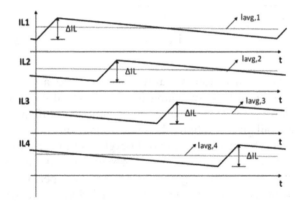

Fig. 10. Inductor current waveforms under steady-state conditions.

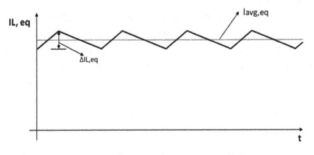

Fig. 11. Equivalent inductor current under steady-state conditions.

Within the time interval [0, DTsw/4] the slope of the equivalent current is given by:

$$S_{eq,on} = \frac{Vin - Vout}{L} - (N-1)\frac{Vout}{L} = \frac{Vin(1 - ND)}{L} \tag{12}$$

The equivalent slope is positive during the first subinterval. During the second subinterval [DTsw/4,Tsw/4] all modules are driven OFF. The equivalent current slope is negative according to the equation:

$$S_{eq,off} = -N\frac{Vout}{L} \tag{13}$$

The equivalent current ripple is given by:

$$\Delta I_{L,eq} = \frac{Vin(1-ND)D}{L}T_{sw} = \frac{\Delta IL\ (1-ND)}{(1-D)} \tag{14}$$

As (14) states, in a multiphase architecture the current ripple is cancelled. If compared with classical buck converters, the current ripple is reduced by (1-ND)/(1-D). Moreover, the equivalent switching frequency as seen by the power load is N times the operating switching frequency of each power module.
Note that (14) could be expressed as:

$$\Delta I_{L,eq} = \frac{Vin(1-ND)D}{L}T_{sw} = \frac{Vin(1-ND)DN\cdot N}{N\cdot NL}T_{sw} = \frac{Vin,eq(1-Deq)Deq}{Leq}T_{sw,eq} \tag{15}$$

where N is the number of modules and $Deq = N\cdot D$, $fsw,eq = N\cdot fsw$, $Leq = \frac{L}{N}$, $Vin,eq = \frac{Vin}{N}$.

By comparing (15) and (5), conclusions are drawn: a N-modules architecture is equivalent to a buck converter operating at $f_{sw,eq}$ switching frequency, $D_{eq}$ duty-cycle, supplied by $V_{in,eq}$ input voltage and featuring $L_{eq}$ inductance. Then, if compared with a buck converter fed by the same input voltage rail and supplying the same voltage bus, the multiphase architecture features higher switching frequency, higher duty-cycle and lower inductance. Both steady-state and dynamic performances of the switching regulator take a great advantage from these interesting properties. With regards to steady-state performances, tight voltage regulation could be achieved thanks to the ripple cancellation effect. Dynamic performances are greatly improved by the reduction of the equivalent inductor under load transients. The steady-state and dynamic performances of multiphase converters are affected by the duty cycle value. A narrow duty cycle results in a poor ripple cancellation. Hence, a larger inductance should be selected to keep tight voltage regulation worsening the transient response. If a wide input-output voltage range is considered, a proper design of passive components is not yet suitable for speeding up the system performances. Multiphase architectures achieving a duty cycle extension by the means of coupled inductors are frequently proposed in literature.
Otherwise, unconventional control algorithms are proposed to speed up the transient response not affecting the passive components design.

## 3. The control system

A survey of the control algorithm is given to highlight the system complexity and high accuracy of simulation results. Authors suggest (Boscaino et al., 2009) for further details. A linear-non-linear digital controller oriented to IC implementation, improving both steady-state and transient response is presented. A schematic block diagram of the linear loop is shown in Fig.12.
A Pulse Width Modulation (PWM) current-mode control is implemented. Two 10 bits/2V ADCs and a 12 bits /5V DAC are included in the analog-digital interface. The output voltage is converted to digital and compared to a reference voltage $V_{ref}$. The error signal

COMP_OUT is generated by the error amplifier which includes the adder and the COMP block. A type-2 compensation action is digitally implemented to ensure system stability. The error signal is then converted to analog and compared to the current sense signal of each power module Vc2,i.

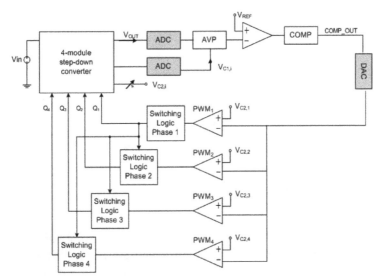

Fig. 12. Schematic block diagram of the linear controller.

The gate drive interleaved technique is digitally implemented by the *Switching logic* subsystems. The master module is the Phase 1, the others slave. Fig.13 shows the switching logic algorithm for the master power module. A clock signal at the switching frequency is synchronously derived from the system clock. The switch-on of the high-side MOSFET is triggered by the rising-edge of the switching clock Ck1. The high-side MOSFET switch-off is triggered by the rising-edge of the corresponding PWM signal.

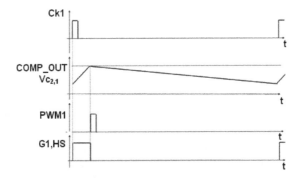

Fig. 13. Switching logic for the master module.

A digital counter generates a switching clock for each slave module which is synchronous with the master switching clock and phase-shifted according to the interleaved technique. The sense signal $V_{c2,i}$ reproduces the instantaneous inductor current of the *i-th* power module. RC networks are used to sense the instantaneous inductor current of each power module. As shown in Fig.14, a RC filter is connected across the low-side MOSFET of each power module. The capacitor voltage $V_{c2}$ is the sense signal for current-mode control implementation.

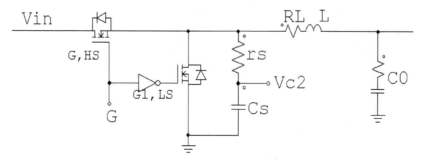

Fig. 14. Current sense network.

The period $T=r_sC_s$ is fixed according to the equation:

$$D_{max}T_{sw} \ll T \tag{16}$$

where $T_{sw}$ is the switching period and $D_{max}$ the maximum operating duty-cycle. The charge and discharge of the capacitor $C_2$ within the switching period reproduces the charge and discharge of the inductor L. The average value of the sense signal is given by:

$$\overline{V_{c2}} = Vout + \overline{I_L}R_L \tag{17}$$

where $\overline{I_L}$ is the average inductor current and $R_L$ the ESR of the inductor L.
The superimposed ripple is given by:

$$\Delta V_{c2} = R_v\Delta I_L \tag{18}$$

where $\Delta I_L$ is the inductor current ripple and $R_v$ the virtual sensing resistor which is given by:

$$R_v = \frac{L}{r_sC_s} \tag{19}$$

The current-mode control is implemented by the Vc2 sense signal. A detailed analysis of the proposed current-sensing technique, including design guidelines is reported in (Boscaino et al., 2008).
Another RC network featuring a lower cut-off frequency is connected across the switching node of the master phase. The sense capacitor voltage $V_{c1}$ reproduces the average inductor current. $V_{c1}$ is converted to digital and feeds the $AVP$ subsystem which implements the well-known Adaptive Voltage Positioning (AVP) technique. Fig.15 shows the output voltage and the load current waveforms under a load current step without AVP. The maximum tolerance window on the output voltage is given by:

$$\Delta V_{out,max} = 2 \cdot \Delta I_{o,max}Rc \tag{20}$$

With the AVP technique, the reference voltage becomes a function of the load current according to the following equation.

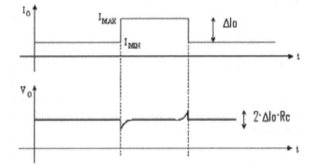

Fig. 15. Dynamic response under load transients. At the top the load current step, at the bottom, the output voltage waveforms are shown.

$$v'_{ref} = V_{ref} - I_o Rc \tag{21}$$

where, $v'_{ref}$ is the variable reference voltage, $V_{ref}$ is a fixed reference voltage, $I_o$ is the load current and $R_c$ is the equivalent series resistor of the output capacitor. The DC set point of the output voltage changes accordingly to the output current, as shown in Fig.16. At steady-state, the output voltage is forced by the AVP technique to the variable reference voltage, instead of a fixed value $V_{ref}$. Under high-load conditions, the DC set point drops in the lower part of the allowable tolerance window. The maximum tolerance window is now given by:

$$\Delta V_{out,max} = \Delta I_{o,max} Rc \tag{22}$$

The technique results in added margin to handle load transients. Since the load current is equal to the average inductor current, the AVP technique is implemented by sensing the average inductor current of the master phase.

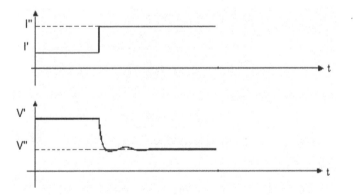

Fig. 16. The dynamic response to a load current step with Adaptive Voltage Positioning.

A non-linear control is introduced to speed-up the transient response not affecting steady-state performances. During transients, the non-linear control modifies a few linear loop parameters, as the reference voltage or the duty-cycle. Since the converter output voltage is the only non-linear loop input, the proposed technique can be efficiently applied to each power supply

system independently of the specific converter topology and linear loop architecture. Thanks to the AVP technique, load transients are detected by an instantaneous drop of the output voltage, and the increasing or decreasing load current transient is detected by monitoring the output voltage drop in terms of sign and width. Two different non-linear control actions are implemented: under a load current increase (*case a*) or decrease (*case b*). The reference voltage (*case a*) or the duty-cycle (*case b*) is modified by the non-linear control loop. Under increasing load current steps, the non-linear control loop acts on the reference voltage advancing the linear loop action. Two auxiliary signals, $V_{VAR}$ and $V_{LOOP}$, are defined: $V_{VAR}$ is the output voltage value related to the new steady-state condition while $V_{LOOP}$ is the output voltage ripple during transient referred to $V_{VAR}$ value, as shown in Fig.17.

$$V_{LOOP} = Vout - V_{VAR} \tag{23}$$

$$v'_{ref} = V_{ref} - V_{LOOP} \tag{24}$$

In order to achieve the fastest transient response, the gain factor $K_{NL}$ is introduced. Hence, $v'_{REF}$ is given by:

$$v'_{ref} = V_{ref} - K_{NL}V_{LOOP} \tag{25}$$

The $K_{NL}$ value is designed by the frequency response analysis to ensure the fastest effective gain under transients avoiding loop instabilities. Under transients, the difference between the output voltage and $V_{VAR}$ is monitored and compared with the maximum steady-state output voltage ripple $V_{OUT\_STEADY}$. If the condition (26) is continuously verified over $T_{STEADY}$ time, the non-linear control action ends.

$$Vout - V_{VAR} < \Delta V_{OUT\_STEADY} \tag{26}$$

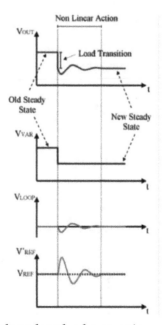

Fig. 17. Non linear control signals under a load current increase.

Under a load current decrease, the non-linear control loop modifies the duty-cycle signal, as shown in Fig.18. All modules are simultaneously turned off to allow the inductor current to decrease as fast as possible. Turning off all modules allows the fastest transient response. Unfortunately, the system is driven towards a loop upsetting and thus to an increase, instead of a reduction, of the output voltage over-and-under shoots. Then, a fixed duty-cycle is forced to all modules according to the interleaved technique. Known the $V_{VAR}$ value, the new steady-state duty-cycle is evaluated and forced by the non-linear subsystem to maintain a regulated output voltage during linear loop resettling. The false steady-state period is a tunable parameter to achieve high design flexibility.

$$D_{NL} = \frac{Vout + \overline{I_L}R_L}{Vin} \tag{27}$$

where Vout is the regulated output voltage, $\overline{I_L}$ is the average inductor current of a single module, $R_L$ the inductor parasitic resistance and Vin the converter input voltage.
From (27), defining the $K_R$ constant as:

$$K_R = \frac{4R_L}{Rc} \tag{28}$$

the forced $D_{NL}$ value is obtained:

$$D_{NL} = \frac{V_{VAR} + (Vout_{I=0} - V_{VAR})K_R}{Vin} \tag{29}$$

The duty-cycle value is evaluated by the non linear controller by the means of fixed parameters and the $V_{VAR}$ value which is obtained by monitoring the output voltage. The difference between the output voltage and $V_{VAR}$ is monitored and compared to the maximum steady-state output voltage ripple $V_{OUT\_STEADY}$. If the condition (30) is continuously verified over a fixed $T_{FSS}$ time, the non-linear control action ends.

$$Vout - V_{VAR} < \Delta V_{OUT\_STEADY} \tag{30}$$

Further design guidelines could be found in (Boscaino et al. 2009, 2010).

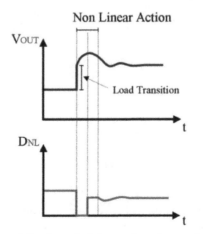

Fig. 18. The output voltage and the forced duty-cycle value under a load current decrease.

## 4. Co-simulation set-up

Active-HDL provides an interface to MATLAB/Simulink simulation environment which allows co-simulation of functional blocks described by using hardware description languages. The digital controller is tested on a FPGA device. A specific co-simulation procedure should be activated in order to make each VHDL design unit available in Simulink libraries. The VHDL code and signal timing are tested as well. High-accuracy simulation results are obtained heavily reducing experimental fault conditions. The co-simulation procedure features automatic data type conversion between Simulink and Active HDL and simulation results could be displayed in both simulation environments. All VHDL entities are collected in an Active HDL workspace. A MATLAB m-file is generated for each VHDL entity. By right clicking on a VHDL entity, select the option "Generate Block Description for Simulink". The procedure is repeated for each VHDL entity of current Workspace. After generating the m-files, the *Active-HDL Blockset* is available in the Simulink library browser. The *Active-HDL Co-Sim* and *HDL Black-Box* elements are used for system co-simulation. The *Active-HDL Co-Sim* element must be included in the Simulink top level model. The block allows the designer to fix co-simulation parameters. In order to simulate a VHDL entity as an integral part of the current model, a *HDL Black-Box* block should be added and associated to the corresponding m-file by the configuration box. When the model is completed, running simulation from Simulink starts the co-simulation. The connection to the Active-HDL server will be opened and all *HDL Black-Boxes* simulated with the Active-HDL simulator. Input and output waveforms can be analyzed both in Simulink and Aldec environment while internal signal waveforms can be analyzed in Aldec environment only.

## 5. MATLAB/Simulink model

The top level model is shown in Fig.19. The power section includes the *Multiphase converter* and the *Current-sense* subsystems. In the *A/D converter* subsystem, the Analog-to-Digital interface is modelled by Simulink library elements. The controller is modelled in Aldec Active-HDL environment and each design unit is included in the *Controller* subsystem. The *Active-HDL Co-Sim* block is added to the top level model.

### 5.1 The power section model

The power section includes the multiphase converter and current-sense filters. A four-module interleaved synchronous buck converter is connected to the power load. The power section is modelled by Simulink library elements in order to achieve the highest processing speed. The use of several co-simulation tools is avoided, keeping at the same time the highest accuracy. In this paragraph, the Simulink model of the multiphase converter based on a continuous-time large-signal model is described. The proposed modelling technique could be efficiently extended to each switching converter.

In Simulink environment, the State-Space block models a system by its own state-space equations:

$$\dot{x} = Ax + Bu$$
$$y = Cx + Du \tag{31}$$

where **x** is the state vector, **u** the input vector and **y** the output vector.

According to (31), the block is fed by the input vector $u$ and the output vector $y$ is generated. As shown in Fig.20, the designer could enter the matrix coefficients in the parameter box.

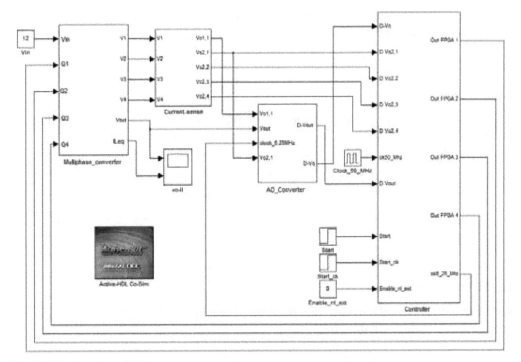

Fig. 19. Top level model

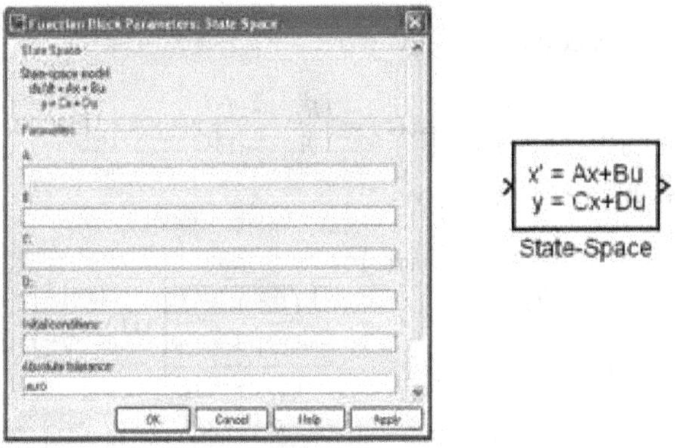

Fig. 20. The State-Space block and the parameter box.

A continuous large-signal model of the buck converter is derived and implemented by the state-space block. The Parasitic components of passive elements as the DC resistance of the inductor (DCR) and the capacitor equivalent series resistor (Rc) are included. Assuming identical MOSFETs, the model accounts for the MOSFET equivalent resistance $R_{ds,on}$. Define a state vector as:

$$x(t) = \begin{bmatrix} x_1(t) \\ x_2(t) \end{bmatrix} = \begin{bmatrix} i_L(t) \\ v_c(t) \end{bmatrix} \tag{32}$$

where $i_L(t)$ is the inductor current and $v_c(t)$ is the voltage across the output capacitor. In the proposed model, the output vector is defined as:

$$y(t) = \begin{bmatrix} y_1(t) \\ y_2(t) \end{bmatrix} = \begin{bmatrix} i_L(t) \\ v_{out}(t) \end{bmatrix} \tag{33}$$

where $v_{out}(t)$ is the output voltage and $i_L(t)$ the inductor current.
During the on-time, the buck converter is equivalent to the circuit shown in Fig.21.

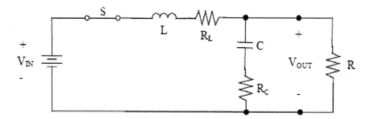

Fig. 21. On-time equivalent circuit.

Both the inductor DCR and the high-side $R_{ds,on}$ are included in the resistance $R_L$. The analysis of the on-time circuits yields:

$$\begin{cases} L\dot{x}_1(t) + R_cC\dot{x}_2(t) = -R_Lx_1(t) - x_2(t) + Vin \\ \\ (Rc + R)C\dot{x}_2(t) = Rx_1(t) - x_2(t) \end{cases} \tag{34}$$

The corresponding matrix equation is given by:

$$\begin{bmatrix} L & RcC \\ 0 & (Rc + R)C \end{bmatrix} \dot{x}(t) = \begin{bmatrix} R_L & -1 \\ R & -1 \end{bmatrix} x(t) + \begin{bmatrix} 1 \\ 0 \end{bmatrix} Vin \tag{35}$$

From (35), the (36) is obtained:

$$\dot{x}(t) = \begin{bmatrix} \dfrac{R_LRc + R(R_L + Rc)}{(Rc + R)L} & -\dfrac{R}{(Rc + R)L} \\ \dfrac{R}{(Rc + R)C} & -\dfrac{1}{(Rc + R)C} \end{bmatrix} x(t) + \begin{bmatrix} \dfrac{1}{L} \\ 0 \end{bmatrix} Vin \tag{36}$$

The OFF-time equivalent circuit is shown in Fig.22. Both inductor DCR and low-side $R_{ds,on}$ are included in $R_L$ resistance.

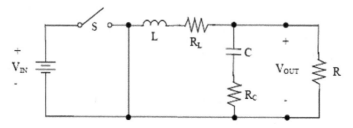

Fig. 22. OFF-time equivalent circuit

The analysis yields:

$$\begin{cases} L\dot{x}_1(t) + RcC\dot{x}_2(t) \doteq -R_L x_1(t) - x_2(t) \\ \\ (Rc + R)C\dot{x}_2(t) = Rx_1(t) - x_2(t) \end{cases} \tag{37}$$

The corresponding matrix equation is given by:

$$\begin{bmatrix} L & RcC \\ 0 & (Rc + R)C \end{bmatrix} \dot{x}(t) = \begin{bmatrix} R_L & -1 \\ R & -1 \end{bmatrix} x(t) \tag{38}$$

Manipulating (38) the (39) is obtained:

$$\dot{x}(t) = \begin{bmatrix} \dfrac{R_L Rc + R(R_L + Rc)}{(Rc + R)L} & -\dfrac{R}{(Rc + R)L} \\ \dfrac{R}{(Rc + R)C} & -\dfrac{1}{(Rc + R)C} \end{bmatrix} x(t) \tag{39}$$

With regard to the output vector, the analysis of both circuits yields:

$$\begin{cases} y_1(t) = x_1(t) \\ y_2(t) = \dfrac{RcR}{R + Rc} x_1(t) + \dfrac{R}{R + Rc} x_2(t) \end{cases} \tag{40}$$

The corresponding matrix equation is given by:

$$y(t) = \begin{bmatrix} 1 & 0 \\ \dfrac{RRc}{R + Rc} & \dfrac{R}{R + Rc} \end{bmatrix} x(t) \tag{41}$$

The analysis of equivalent circuits results in two linear state-space models. During the on-time, the converter is modelled by state-space equations (42):

$$\dot{x}(t) = A_{on} \cdot \dot{x}(t) + b_{on} \cdot u(t)$$

$$y(t) = C_{on} \cdot \dot{x}(t) + d_{on} \cdot u(t) \tag{42}$$

where:

$$A_{on} = \begin{bmatrix} \dfrac{R_L Rc + R(R_L + Rc)}{(Rc + R)L} & -\dfrac{R}{(Rc + R)L} \\ \dfrac{R}{(Rc + R)C} & -\dfrac{1}{(Rc + R)C} \end{bmatrix} \tag{43}$$

$$b_{on} = \begin{bmatrix} \dfrac{1}{L} \\ 0 \end{bmatrix} \tag{44}$$

$$C_{on} = \begin{bmatrix} 1 & 0 \\ \dfrac{RRc}{R + Rc} & \dfrac{R}{R + Rc} \end{bmatrix} \tag{45}$$

$$\mathbf{d_{on}} = \begin{bmatrix} 0 \\ 0 \end{bmatrix} \tag{46}$$

During the OFF-time, the converter is modelled as:

$$\mathbf{x(t)} = \mathbf{A_{off}} \cdot \mathbf{x(t)} + \mathbf{b_{off}} \cdot u(t)$$

$$\mathbf{y(t)} = \mathbf{C_{off}} \cdot \mathbf{x(t)} + \mathbf{d_{off}} \cdot u(t) \tag{47}$$

where

$$\mathbf{A_{on}} = \begin{bmatrix} \dfrac{R_L Rc + R(R_L + Rc)}{(Rc + R)L} & -\dfrac{R}{(Rc + R)L} \\ \dfrac{R}{(Rc + R)C} & -\dfrac{1}{(Rc + R)C} \end{bmatrix} \tag{48}$$

$$\mathbf{b_{off}} = \begin{bmatrix} 0 \\ 0 \end{bmatrix} \tag{49}$$

$$\mathbf{C_{off}} = \begin{bmatrix} \dfrac{1}{RRc} & 0 \\ \dfrac{1}{R + Rc} & \dfrac{R}{R + Rc} \end{bmatrix} \tag{50}$$

$$\mathbf{d_{off}} = \begin{bmatrix} 0 \\ 0 \end{bmatrix} \tag{51}$$

Note that:

$$\mathbf{A} = \mathbf{A_{ON}} = \mathbf{A_{OFF}}$$

$$\mathbf{C} = \mathbf{C_{ON}} = \mathbf{C_{OFF}}$$

$$\mathbf{b_{OFF}} = 0$$

$$\mathbf{d_{ON}} = \mathbf{d_{OFF}} = 0 \tag{52}$$

Define the gate drive signal $q(t)$ as:

$$q(t) = \begin{cases} 1 \; during \; the \; ON \; time \\ 0 \; during \; the \; OFF \; time \end{cases} \tag{53}$$

With the aid of q signal, a new input signal is defined as:

$$u(t) = q(t) \cdot Vin \tag{54}$$

A unique state-space model is thus obtained:

$$\mathbf{x(t)} = \mathbf{A} \cdot \mathbf{x(t)} + \mathbf{b} \cdot Vin \cdot q(t)$$

$$\mathbf{y(t)} = \mathbf{C} \cdot \mathbf{x(t)} \tag{55}$$

The Simulink model of the buck converter is shown in Fig.23.

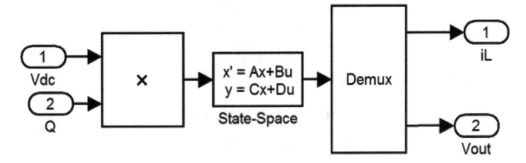

Fig. 23. The buck converter model.

In the state space block, matrix coefficients are entered. The input voltage $V_{dc}$ and the gate drive logic signal Q are the subsystem inputs. The scalar components of the output vector are obtained by the *Demux* block.

The modelling technique is extended to the four-modules architecture which is shown in Fig.24.

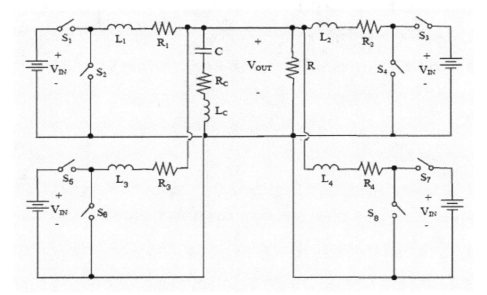

Fig. 24. A four-modules buck converter.

The state-vector is defined as:

$$x(t) = \begin{bmatrix} i_{L1}(t) \\ i_{L2}(t) \\ i_{L3}(t) \\ i_{L4}(t) \\ v_c(t) \end{bmatrix} \tag{56}$$

The output vector is defined as:

$$y(t) = \begin{bmatrix} i_{L1}(t) \\ i_{L2}(t) \\ i_{L3}(t) \\ i_{L4}(t) \\ v_{out}(t) \end{bmatrix} \tag{57}$$

The state-space model is obtained:

$$\dot{x}(t) = A \cdot x(t) + b \cdot Vin \cdot q(t)$$

$$y(t) = \dot{C} \cdot x(t) \tag{58}$$

where:

$$A = \begin{bmatrix} \dfrac{RR_1 + RR_C + R_C R_1}{L_1(R + R_C)} & -\dfrac{RR_C}{L_1(R + R_C)} & -\dfrac{RR_C}{L_1(R + R_C)} & -\dfrac{RR_C}{L_1(R + R_C)} & -\dfrac{R}{L_1(R + R_C)} \\[2mm] -\dfrac{RR_C}{L_2(R + R_C)} & -\dfrac{RR_2 + RR_C + R_C R_2}{L_2(R + R_C)} & -\dfrac{RR_C}{L_2(R + R_C)} & -\dfrac{RR_C}{L_2(R + R_C)} & -\dfrac{R}{L_2(R + R_C)} \\[2mm] -\dfrac{RR_C}{L_3(R + R_C)} & -\dfrac{RR_C}{L_3(R + R_C)} & -\dfrac{RR_3 + RR_C + R_C R_3}{L_3(R + R_C)} & -\dfrac{RR_C}{L_3(R + R_C)} & -\dfrac{R}{L_3(R + R_C)} \\[2mm] -\dfrac{RR_C}{L_4(R + R_C)} & -\dfrac{RR_C}{L_4(R + R_C)} & -\dfrac{RR_C}{L_4(R + R_C)} & -\dfrac{RR_4 + RR_C + R_C R_4}{L_4(R + R_C)} & -\dfrac{R}{L_4(R + R_C)} \\[2mm] \dfrac{R}{C(R + R_C)} & \dfrac{R}{C(R + R_C)} & \dfrac{R}{C(R + R_C)} & \dfrac{R}{C(R + R_C)} & -\dfrac{1}{C(R + R_C)} \end{bmatrix} \tag{59}$$

$$B = \begin{bmatrix} \dfrac{1}{L_1} & 0 & 0 & 0 \\[2mm] 0 & \dfrac{1}{L_2} & 0 & 0 \\[2mm] 0 & 0 & \dfrac{1}{L_3} & 0 \\[2mm] 0 & 0 & 0 & \dfrac{1}{L_4} \end{bmatrix} \tag{60}$$

$$C = \begin{bmatrix} 1 & 0 & 0 & 0 & 0 \\ 0 & 1 & 0 & 0 & 0 \\ 0 & 0 & 1 & 0 & 0 \\ 0 & 0 & 0 & 1 & 0 \\ \dfrac{RR_C}{R + R_C} & \dfrac{RR_C}{R + R_C} & \dfrac{RR_C}{R + R_C} & \dfrac{RR_C}{R + R_C} & \dfrac{R}{R + R_C} \end{bmatrix} \tag{61}$$

$$q(t) = \begin{bmatrix} q_1(t) \\ q_2(t) \\ q_3(t) \\ q_4(t) \end{bmatrix} \tag{62}$$

Fig. 25 shows the Simulink model of the multiphase converter. The input voltage and scalar components of the q vector are the model inputs. The *Mux* block multiplexes scalar components into the input vector **u**. The *Demux* block generates the scalar components of the output vector **y**(t).

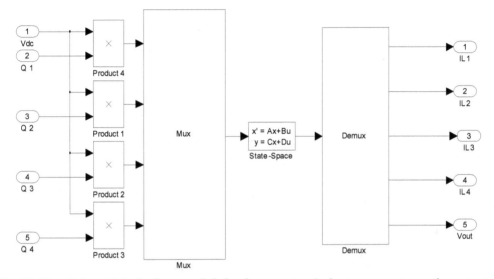

Fig. 25. Simulink model of a four-module buck converter. Inductor currents are the output vector components.

The proposed model could be adapted to the designer requirements. For example, in Fig.26 another implementation is shown. Besides the inductor current signals and the output voltage, the switching node voltages $V_{sw}$ of each module are also included in the output vector. Further, the equivalent inductor current instead of the inductor currents as well as switching node voltages are available outside the subsystem. The equivalent current is obtained as the sum of inductor currents by the means of *Adder* blocks.

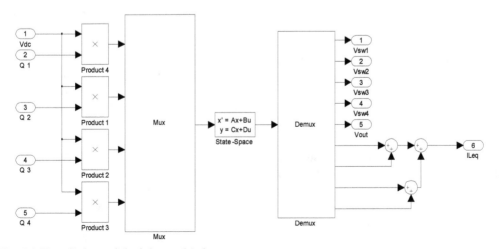

Fig. 26. Simulink model of the multiphase converter.

## 5.2 The sense network

Current sense filters are modelled by their own Laplace transfer function using the *Transfer Function* block. In the parameter box the coefficients of the Laplace function are entered. The *low-pass filter* subsystem of the master phase is shown in Fig.27. Low-pass filter subsystems are fed by the switching node voltage of the corresponding power module.

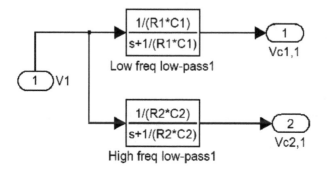

Fig. 27. Current-sense filters of the master module.

## 5.3 The digital controller model

The digital controller is modelled as the collection of VHDL entities, each one described by the VHDL code in Active-HDL environment. By the co-simulation procedure, each design unit is added to the Simulink model and simulated among the others. The Simulink model of the *Controller* subsystem is shown in Fig.28. Two subsystems are highlighted: the *Non_linear_controller* and the *Linear_Controller*. VHDL entities are collected in the *Controller* subsystem. The 6.25MHz system clock is derived from the 50MHz clock signal available from the FPGA device by the means of a digital clock divider, described in the *clock_div* entity. VHDL entities are linked in Simulink environment. Note that modelling the digital controller by multiple entities is a unique choice of the designer. Instead of modelling the controller by a unique VHDL entity, authors suggest a multiple entities approach to monitor internal signal of the FPGA controller in Simulink environment. The co-simulation procedure allows the designer to test the VHDL code, including timing between entities. The digital controller is thus tested as closely as possible to the effective FPGA implementation.

Each VHDL entity accomplishes a specific task of the digital controller: soft-start, protection circuits for safe operation, adaptive voltage positioning and the strictly-named controller implementing the sophisticated control algorithm, as shown in Fig.29. The co-simulation procedure as well as the proposed modelling approach allows the designer to test each available function by behavioural simulations, thus reducing the risk of experimental fault conditions.

In the *Linear Controller*, the error signal is converted to analog by the DAC subsystem and then compared to the analog saw-tooth within the PWM generator. The interleaved technique is digitally implemented. Fig.30 shows the switching logic subsystem of the master phase. The *f1_sim* VHDL entity is fed by the master phase PWM signal and non-linear control signals. The unit generates the q-signal for the master module. The digital counter which generates the switching clocks (*SwCk*) of all slave modules is modelled by the *acc_sint* VHDL unit.

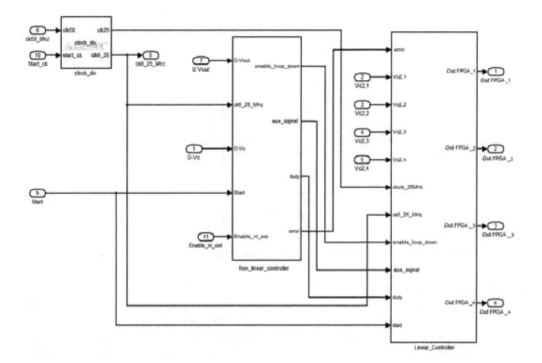

Fig. 28. The Controller subsystem.

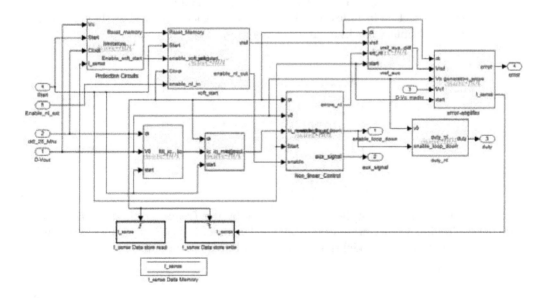
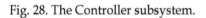

Fig. 29. The non-linear controller.

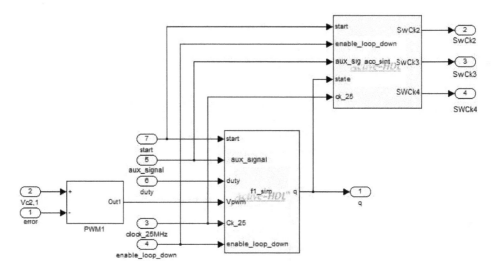

Fig. 30. The Switching Logic model of the master module.

## 5.4 Analog-digital interface

Co-simulation features are to be well-understood before implementing a mixed-mode model. ALDEC and Simulink automatically code common signals. In ALDEC environment signals are coded as signed binary vectors. The output of each HDL block is thus a binary vector. Simulink automatically codes binary vectors by the equivalent decimal number. HDL block inputs are Simulink signals thus coded as decimal numbers. ALDEC environment codes decimal numbers as the equivalent signed binary vector. The Simulink model should account for binary-decimal data conversion. Digital to analog converters and analog to digital converters are introduced in the Simulink model.

The A/D converter section includes two A/D converters, each modelled as shown in Fig.31 by the means of elementary blocks of Simulink libraries.

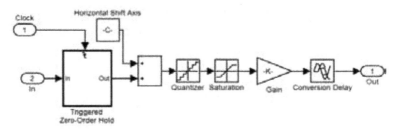

Fig. 31. The A/D converter model

The input signal is sampled at the clock frequency by the *Triggered Zero-Order Hold* block. Since in the Simulink library browser only an ideal symmetrical *Quantizer* block is provided, a non-ideal A/D converter model is achieved by adding the horizontal shift. The horizontal shift constant is given by:

$$C = \frac{FS}{2^{Nbit+1}}$$

(63)

where $N_{bit}$ is the bit resolution and FS the full scale voltage.

The available *Quantizer* block allows the user to set the voltage resolution while the number of levels is yet unlimited.

The *Saturation* block is thus cascaded to the *Quantizer* in order to set the full scale range. The high-level saturation limit is given by:

$$\text{High} = V_{res} \cdot N_{lev} = \frac{FS}{2^{Nbit}} \cdot (2^{Nbit} - 1) \tag{64}$$

where $V_{res}$ is the voltage resolution and $N_{lev}$ the number of levels. The discrete output of the *Saturation* block should be coded. In order to simulate the ADC coding, the *Gain* block is cascaded to the saturation block. The output of the ADC model is the equivalent decimal number of the binary vector. The conversion gain is given by:

$$K = \frac{FS}{2^{Nbit}} \tag{65}$$

The *Conversion Delay* block accounts for the input-output delay of the modelled A/D converter.

The Simulink model of the digital-to-analog converter is shown in Fig.32. The subsystem is fed by a digital signal and generates a discrete analog signal. The maximum value of the input signal is given by:

$$Vmax = 2^{N_{bit,dac}} - 1 \tag{66}$$

where $N_{bit,dac}$ is the converter bit resolution. The maximum value of the output voltage is equal to the DAC full scale voltage. The digital-to-analog converter is modelled by a gain factor $K_{dac}$:

$$K_{dac} = \frac{FS_{dac}}{2^{N_{bit,dac}}} \tag{67}$$

In order to model the full scale range of the converter, a *Saturation* block is cascaded with the gain block. Saturation levels are equal to the full scale range.

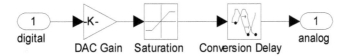

Fig. 32. The Simulink model of a digital-to-analog converter.

## 5.5 PWM comparator model

The PWM comparator is fed by the analog error signal and the analog current sense signal. The analog comparator model is shown in Fig.33. An adder block is used as the input stage of the comparator. The *Relay* block models the non-linearity of the comparator. The Relay block allows its output to switch between two specified values. The state of the relay is obtained by comparing the input signal to the specified thresholds, the *Switch-off point and the Switch-on point* parameters. When the relay is on, it remains on until the input drops below the value of the *Switch-off point* parameter. When the relay is off, it remains off until the input exceeds the value of the *Switch-on point* parameter. The parameter box allows the end user to set a specified value for both on and off conditions. In this model, the output value is equal to "1" in the on-state, "0" in the off-state.

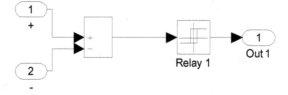

Fig. 33. The Simulink model of the PWM comparator.

## 6. Simulation results

A 12V/1V@120A four-module buck converter has been designed and modelled in Simulink-Aldec mixed environment. Each power module operates at 250kHz switching frequency. Simulation results are shown to highlight the high accuracy of the modelling approach. Steady-state waveforms under 1A load are shown in Fig.34. At the top the output voltage waveform, at the bottom the equivalent inductor current is shown. The proposed modelling approach allows the designer to simulate the system behaviour within the switching period achieving high accuracy results. The output voltage ripple and the current ripple are accurately modelled. The DC set point of the output voltage is fixed at the nominal value. The output voltage, at the top, and the equivalent current, at the bottom, under 115A load are shown in Fig.35. The DC set point is fixed at 0.96V at full load due to the implemented AVP technique.

The system is tested under load current transients to evaluate the non-linear control performances. Under a 0-120A load current step, if the non linear controller is disabled the system behaviour shown in Fig.36 is obtained. Under a 120A -0 load current step, if the non linear controller is disabled the system behaviour shown in Fig.37 is obtained. The output voltage, at the top, and the equivalent current, at the bottom, are shown. Note that the proposed design approach accurately models the system behaviour within the switching period as well as under large-signal transients.

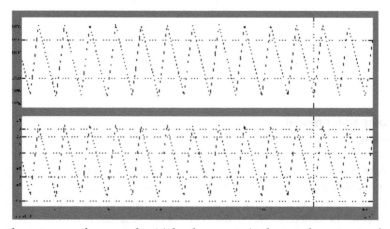

Fig. 34. Steady-state waveforms under 1A load current. At the top the output voltage, at the bottom the equivalent current waveforms are shown.

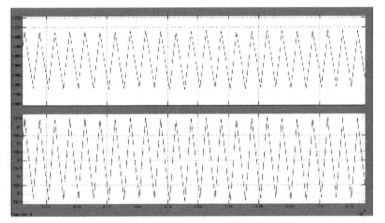

Fig. 35. Steady-state waveforms under 115A load current. At the top the output voltage, at the bottom the equivalent current waveforms are shown.

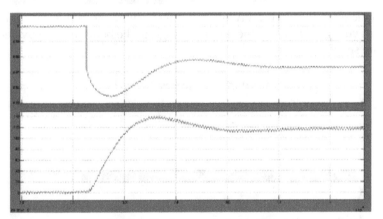

Fig. 36. Dynamic response to a 0-120A load current step without the non-linear control. At the top the output voltage, at the bottom the equivalent current waveforms are shown.

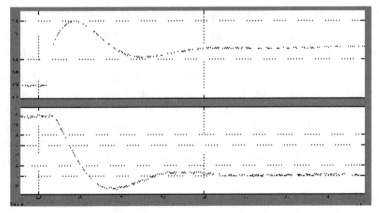

Fig. 37. Dynamic response to a 120A-0 load current step without the non-linear control. At the top the output voltage, at the bottom the equivalent current waveforms are shown.

Fig.38 and 39 show the dynamic response to a 0-120A and 120A-0 load current step, respectively, obtained by enabling the non-linear controller.

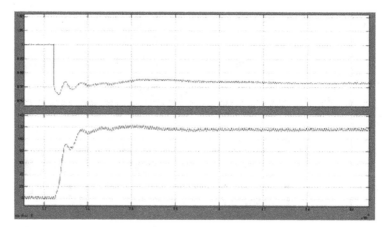

Fig. 38. Dynamic response to a 0-120A load current step with the non-linear control. At the top, the output voltage. At the top the output voltage, at the bottom the equivalent current waveforms are shown.

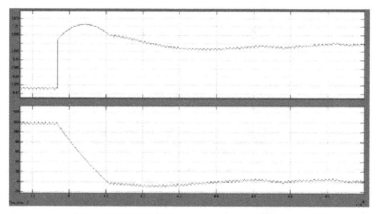

Fig. 39. Dynamic response to a 120A-0 load current step with the non-linear control. At the top the output voltage, at the bottom the equivalent current waveforms are shown.

As shown by simulation results, the non linear controller ensures a reduction of the output voltage over- and-under-shoots during large-signal transients. Under the rising edge of the load current step, the output voltage under-shoot is reduced by 22%. Under the falling edge of the load current step, the output voltage over-shoot is reduced by 26%. The recovery time is reduced by 24%. The instantaneous output voltage change due to the capacitor ESR is included. Note that the instantaneous drop is not affected by the non-linear control.

## 7. Experimental results

A laboratory prototype has been realized to test the efficiency of the non-linear control algorithm. Experimental validation of the non linear control algorithm has been proposed in

(Boscaino et al., 2010). The high accuracy of the proposed design approach is here discussed by the comparison of simulation and experimental results. Experimental waveforms in Fig.40 show the system behaviour under 120A load current square-wave obtained by disabling the non-linear control (a), and by enabling the non-linear control (b). Ch1 shows the output voltage (200mV/div), Ch3 shows the master module inductor current (10A/div) and Ch4 shows the load current (60A/div).

The great potential of the proposed design approach relies in the performance evaluation of the designed controller. Under load transients, the recovery time as well as the percentage of drop reduction due to the non linear control accurately matches experimental results. The system behaviour is perfectly matched by simulation results. Evaluating system performances by simulation or experimental results leads to the same conclusions. A detail of Fig.40 under the rising edge of the load current square wave is shown in Fig.41: (a) the linear controller response, (b) the non-linear controller response. As shown by experimental results, under the load current rising-edge, the under-shoot of the output voltage is reduced by 29%.

A detail of Fig.40 under the falling edge of the load current square wave is shown in Fig.42: at the left the linear controller response, at the right the non-linear controller response. By comparing experimental waveforms in Fig.42 the output voltage over-shoot under a load current falling edge is reduced by 28%. Under both transients, the recovery time is reduced by 25%.

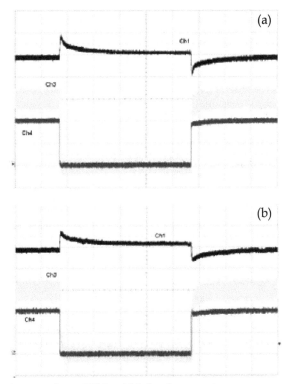

Fig. 40. System behaviour under a 115A width load current square wave. Ch1 shows the output voltage (200mV/div), Ch3 shows the master module inductor current (10A/div) and Ch4 shows the load current (60A/div). (a) The non-linear control is disabled. (b) The non-linear controller is enabled.

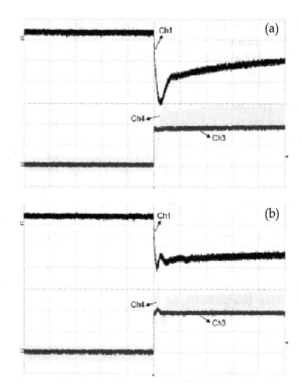

Fig. 41. System behaviour under the load current rising-edge. Ch1 shows the output voltage (50mV/div, AC coupling), Ch3 shows the master module inductor current (10A/div) and Ch4 shows the load current (60A/div). At the left, the non-linear control is disabled. At the right, the non-linear controller is enabled.

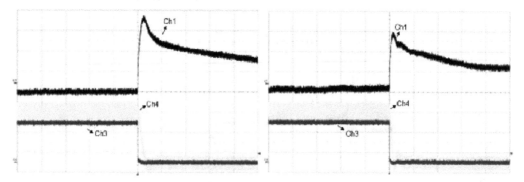

Fig. 42. System behaviour under the load current falling-edge. Ch1 shows the output voltage (50mV/div, AC coupling), Ch3 shows the master module inductor current (10A/div) and Ch4 shows the load current (60A/div). At the left, the non-linear control is disabled. At the right, the non-linear controller is enabled.

## 8. Conclusion

In this chapter, a high-accuracy design approach for power supply systems is proposed. Mixed-mode systems are modelled in Simulink environment using powerful co-simulation tools. For the highest processing speed, a modelling technique for the power section is proposed. The digital subsystems are described by VHDL source code and verified by Aldec/Simulink co-simulation tool, allowing the test of the VHDL code and timing between separate entities. As an application, the model of a multiphase dc-dc converter for VRM applications is described. Thanks to the available co-simulation tool, all provided functions such as soft-start, protection algorithms for safe operation and adaptive voltage positioning are accurately modelled and verified during behavioural simulations. As shown by simulation results, the proposed model matches the system behaviour both within the switching period and during load transients. The design approach allows the designer to simulate the system behaviour as closely as possible to the effective behaviour, modelling and testing the "effective" controller. The controller performances are evaluated by simulation and experimental results leading to the same conclusions. Authors suggest the co-simulation tool for mixed-mode system and complex architectures. Actually, the IC implementation of the proposed controller in BCD6S technology is carried on. A central unit for signal processing is introduced. Mathematical operations are time-multiplexed in order to speed up the input-output delay in the control loop. The co-simulation tool is already used to test the efficiency of the implemented solution.

## 9. Acknowledgements

Authors would like to thank ST Microelectronics for supporting this research activity.

## 10. References

Basso, C. (2008). *Switch-mode Power Supplies*, McGraw Hill, ISBN 978-0-07-150859-0, USA.

Boscaino, V., Livreri, P., Marino, F. & Minieri, M. (2008). Current-Sensing Technique for Current-Mode Controlled Voltage Regulator Modules. *Microelectronics Journal*, Vol.39, No. 12, (December 2008), pp. 1852-1859, ISSN:0026-2692.

Boscaino, V., Livreri, P., Marino, F. & Minieri, M. (2009). Linear-non-linear digital control for dc/dc converters with fast transient response. *International Journal of Power and Energy Systems*, Vol.29, No.1, (March, 2009), pp. 38-47, ISSN: 10783466.

Boscaino, V., Gaeta, Capponi, G. & M., Marino, F. (2010). Non-linear digital control improving transient response: Design and test on a multiphase VRM, *Proceedings of the International Symposium on Power Electronics, Electrical Drives, Automation and Motion*, ISBN: 978-142444987-3, Pisa, Italy, June 2010.

Erickson, R. & Maksimovic, D. (2001). *Fundamentals of Power Electronics (2nd ed)*, Springer Science, ISBN 0-7923-7270-0, USA.

Kaichun, R. et al. (October 2010). Matlab Simulations for Power Factor Correction of Switching Power, In: *Matlab -Modelling, Programming and Simulations*, Emilson Pereira Leite, pp. 151-170, Scyio, ISBN 978-953-307-125-1, Vukovar, Croatia.

Miftakhutdinov, R. (June 2001). Optimal Design of Interleaved Synchronous Buck Converter at High Slew-Rate Load Current Transients. IEEE *Power Electronics Specialists Conference*, ISBN 0-7803-7067-8, Vancouver, BC, Canada, June 2001.

Pop, O. & Lungu, S. (October 2010). Modelling of DC-DC Converters, In: *Matlab -Modelling, Programming and Simulations*, Emilson Pereira Leite, pp. 125-150, Scyio, ISBN 978-953-307-125-1, Vukovar, Croatia.

Pressman, A. & Billings, K. (2009). *Switching Power Supply Design (3rd ed.)*, McGraw Hill, ISBN 978-0-07-148272-1, USA.

# 2

# High Accuracy Modelling of Hybrid Power Supplies

Valeria Boscaino and Giuseppe Capponi
*University of Palermo*
*Italy*

## 1. Introduction

This chapter proposes a modelling approach based on the PSIM/Simulink co-simulation toolbox for hybrid power supplies, featuring high accuracy. Hybrid source performances are fully tested during behavioural simulations. The importance of high-accuracy modelling is investigated and modelling guidelines for both power sources and load are given for further applications. The Simulink model is analyzed and the efficiency of the proposed approach is verified by the comparison between simulation and experimental results.

In the last few years, hybrid power supplies are investigated for a wide variety of application areas: two primary sources are coupled to take advantages of both, overcoming their drawbacks. The growing interest in hybrid sources is mainly due to the spread of fuel cells.

Fuel cells are renewable energy sources, fed by an external fuel and thus potentially infinite source of energy. The fuel cell supplies power until the fuel is supplied to, thus ensuring potentially infinite life cycle to each power load. Automotive, residential and portable electronics are only a few examples of fuel cells application areas. Powering a portable electronic device by a fuel cell is still a challenge for the scientific world. With the aid of fuel cells, the full portability is achieved: electronics devices could be recharged by simply replacing the fuel cartridge instead of being connected to the power grid. Further, fuel cells ensure the highest energy density allowing higher and higher device run time. Yet, portable devices feature a pulsed power consumption profile which depends on the user-selected function. The required peak power is usually higher and higher than the average power. Since the response of fuel cells to instantaneous power demands is relatively poor, innovative solutions are investigated to take full advantage from the fuel cell use. The goal is to couple the fuel cell with a high power density component. The instantaneous peak power is supplied by the high power density component while the average power is supplied by the fuel cell itself. Specific power management algorithms for active power sharing are required. Life-cycle and transient response are directly controlled thus leading to a close dependence of the power management control algorithms on the related application and power load. In the last few years, authors experienced the importance of accurate modelling of both power source and load to evaluate the system performances since the simulation step. The efficiency of power management algorithms is ensured by the high accuracy of source, power load and power management system modelling.

The choice of a powerful simulation environment is a key step for a successful conclusion of the overall design process. The power consumption profile of the specific load device should be deeply analyzed to ensure proper control algorithms. Load emulators best fit designers' requirements. During behavioural simulations, emulators allow the designer to virtually use the load device reproducing the effective power consumption profile. Hence, no assumptions are made on the load power consumption profile while designing the power management algorithms avoiding the risk of experimental failures.

Fuel cell modelling is a critical step since fuel cells are non linear devices whose performances are heavily affected by temperature, fuel pressure and voltage-current operating conditions. Since electro-chemical, physical and geometrical parameters are not readily available to applications engineers, a black-box approach is usually preferred to an analytical one. Both steady-state and dynamic behaviour should be accurately modelled in order to design an efficient power supply design. In this chapter, an accurate modelling approach is proposed.

As an application, a fuel cell – supercapacitor hybrid power supply for portable applications is designed and tested. The proposed system supplies a Digital Still Camera (DSC). Like any other portable device, the digital still camera behaviours as a pulsed load. Unlike the others, the power consumption profile features three possible standby modes (playback, photography and movie mode), each one requiring a specific average current value. The load power consumption profile is closely dependent on the user-selected function and thus unpredictable. Consequently, the DSC represents the worst-case portable device for hybrid power supply design and optimization. Note that accurate modelling is required for testing purposes only. Useful information about the DSC state is reproduced by sampling the supply voltage. The power load, a Fujifilm S5500 Digital Still Camera is modelled in Simulink environment with the aid of the *Stateflow* toolbox. A user-interactive interface is provided. During behavioural simulations, the designer is able to select a specific DSC function and the corresponding power consumption profile is emulated. The fuel cell is modelled in Simulink environment. The architecture and modelling technique is suitable for each fuel cell type and power level. The proposed fuel cell model has been implemented by authors in several simulation environments like PSIM, Simulink and PSPICE and further implementations are readily available in literature. Yet, authors suggest the Simulink implementation to achieve low-complexity, high accuracy and the highest processing speed. On the basis of the Simulink model, a FPGA-based fuel cell emulator has been realized. With the aid of the fuel cell emulator, preliminary tests could be performed saving hydrogen reserve. A circuit implementation of the analog controller in PSIM environment is proposed. PSIM by Powersim Inc. is a circuit simulation software oriented to power electronics systems. Furthermore, a Matlab/PSIM co-simulation tool allows behavioural simulation in a mixed environment. In the basic package, PSIM consists of two programs: PSIM Schematics, full-featured schematic-entry program and SIMVIEW, advanced waveforms viewer software. As soon as the PSIM schematic is built, the designer is able to start the co-simulation procedure. In PSIM Elements/Control Library, the *SimCoupler Module* is provided for Matlab/Simulink co-simulation. Simulink input and output nodes are available. The Simulink input nodes receive values from Simulink and behaviour as voltage sources while the output nodes send values to Simulink. Input and output nodes can be placed in the PSIM schematics according to co-simulation requirements. After generating the circuit netlist in PSIM environment, the PSIM procedure for system co-simulation is completed. In Simulink, the *SimCoupler* block can be added to the existing

Simulink model and linked to the PSIM netlist file by the block parameter box. The *SimCoupler* block now behaviours as a Simulink subsystem showing input and output ports as specified in the corresponding PSIM schematic: the Simulink output nodes are the subsystem input ports and the Simulink input nodes are the subsystem output ports. The PSIM schematic is simulated as an integral part of the active Simulink model. The efficiency of the Simulink-based modelling technique is shown by the comparison between simulation and experimental results.

## 2. Hybrid power supply architectures

Hybrid power sources have received a great deal of attention in the last few years and they are considered the most promising source of energy for portable applications. Two basic sources are coupled: a high energy density and a high power density source. The goal is to optimize the power and energy performances of the composite power source. Each basic source supplies the load current under a specific working condition, as steady-state or transient, according to its own stand-alone performances. The high energy density source, i.e. fuel cell, supplies the load current under steady-state. Simultaneously, under a load transient the high power density component, i.e. battery or supercapacitor, supplies the load with the difference between the instantaneous and the steady-state load current. The controller is accurately designed to correctly manage the power flow between primary sources and load. If properly controlled, the hybrid source is neither influenced by the transient response of the high energy density source nor by the running time of the high power density source. If compared with stand-alone basic sources, the hybrid source shows longer running time, faster transient response, higher power and energy density.

In literature, battery-ultracapacitor or fuel cell-battery power supply systems are presented (Gao &Dougal, 2003; Alotto et al., 2008). Coupling a battery and an ultracapacitor yields higher specific power than a battery-alone source and higher specific energy than an ultracapacitor-alone source. In the meantime higher running time is achieved. Research on battery/ultracapacitor hybrids has been actively carried out for electric vehicles, portable electronics, weapon systems, as well as for industrial applications.

In literature hybrids are classified in active and passive sources. Direct coupling is a passive hybrids feature while in active hybrids a DC-DC converter is interposed between the two basic sources. Several advantages are brought by the DC/DC converter. A detailed comparative analysis of a battery/supercapacitor hybrid in passive and active configuration is reported in (Gao & Dougal, 2003). The active hybrid configuration ensures an increase in power capability, a better regulation of the output voltage, a minimization of the current ripple and a reduction of the system weight and volume. If a capacitor based hybrid is considered, since the voltage of the two basic sources may be different, the capacitor array size could be designed independently of the other primary source. The DC-DC converter plays an important role in regulating the output voltage and making the voltage of the two basic sources independent of each other. The active hybrids ensure higher power capability than passive hybrids since the power sharing is directly controlled by the switching converter. Further, in active hybrids the output voltage is independent of the discharge curve of its stand-alone sources. In battery-based hybrids, the DC/DC converter can also act as a battery charger, leading to a decrease in system cost and weight.

Yet, all described hybrid systems are designed and tested with regular pulsed load, featuring a square wave current profile. The frequency and width of the current profile are usually assumed as design ratings by control algorithms, even if these assumptions are not verified by commercial devices. If tested with commercial devices, these algorithms would lead to an accidental turning-off or alternatively to a weak optimization of the hybrid power source in terms of both running time and power capability. A fuel cell/supercapacitor active hybrid system supplying a digital still camera is presented. A buck converter is interposed between the primary sources. The specific load device affects system ratings and the architecture of the hybrid source but not the control algorithm. Even under stress, an accidental turning-off is avoided. A high-accuracy model is proposed for both power sources and load. The importance of modelling is confirmed by the comparison between simulation and experimental results.

## 3. The importance of measurements

The importance of accurate modelling is extensively analyzed in literature. A detailed analysis of several modelling techniques is reported in (Liu, 1993). Measurement-based models are here proposed for both power source and load. Critical features are accurately highlighted and characterized by well-defined measurements. Designers accurately plan the measurement stage to achieve the best accuracy as well as the lowest complexity. Since the modelling approach can be efficiently applied to other devices, performed measurements are described for further applications. The model accuracy is closely related to the experimental setup. The goal is to minimize the difference between the experimental and simulated response to the same input which is usually referred to as the model error. The least the model error, the more efficient the model is. Under ideal conditions, the error is null and the model perfectly matches the effective behaviour of the device under test (DUT). A source of error relies in the model architecture. For example, a highly non-linear device is usually modelled by a linear model leading to an intrinsic source of error. Further, the model error is affected by the accuracy of measures since the model parameters are estimated from a finite and noisy data set. The proposed models are based on high-accuracy measurements and a non-linear model is proposed despite of the model complexity. Tunable parameters are provided for further applications.

### 3.1 Measurements on the load device
The digital camera behaviours as a pulsed load whose power consumption profile highlights two critical features: the power consumption profile is unpredictable since it depends on the user-selected function and the peak power is usually higher and higher than the average power drawn by the power supply. The power source life cycle and fuel consumption as well as the tolerance window on the output voltage are only a few performance indicators of the hybrid power supply. Hence, a complete analysis is performed to identify critical working conditions. The goal is to reproduce the power consumption profile of the load device and its correlation with the running function. The proposed model allows the designer to virtually use the digital camera during behavioural simulations. The designer is able to select a function obtaining the required consumption profile. The supply current of the digital camera has been monitored under function running. Experimental data are stored in look-up-tables and included in the Simulink model. Samples of measures are here discussed. The DSC features three stand-by modes of operation playback, photography and movie mode. Each mode can be maintained indefinitely active even if no function is running. The standby current is constant and

closely related to the actual mode: 180mA for playback mode, 330mA for photography mode and 440mA for movie mode. Each available function requires a pulsed current over the entire function run time. The load current profile under turning-on in playback mode is shown at the left in Fig.1. At the end of function transient, the required load current is strictly constant at 180mA. Yet, during the transient, the peak current reaches 486mA and the run time is equal to 1.24s. The load current profile under the zoom function in playback mode is shown at the right in Fig.1. The transient time is equal to 88.4ms and the peak current reaches 351mA. The average current value over the function transient time is equal to 269mA. If compared with turning-on in playback mode, the transient time is quite short and the peak current value is quite similar to the average current value. Hence, the zoom function is not critical for the hybrid power supply.

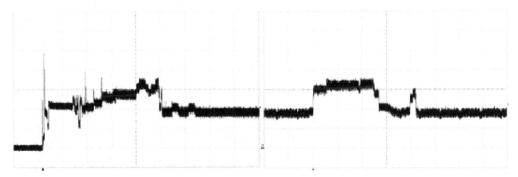

Fig. 1. At the left, the DSC current profile under turning-on in playback mode (100mA/div) is shown. Time base is set to 200ms/div. At the right, the DSC current profile under zoom function in playback mode (100mA/div) is shown. Time base is set at 20ms/div.

The load current profile under turning-on the DSC in photography mode is shown at the left in Fig.2. At the end of transient the current is constant at 330mA which is the standby current of the photography mode. Yet, under function running, the load current reaches a 611mA peak value and the transient time is equal to 4.6s. The load current profile required for taking picture without flash is shown at the right in Fig.2. The transient time is equal to 2.9s. The peak current value is equal to 716mA and the average current over the transient time is equal to 443mA. The load current under movie mode selection is shown at the left in

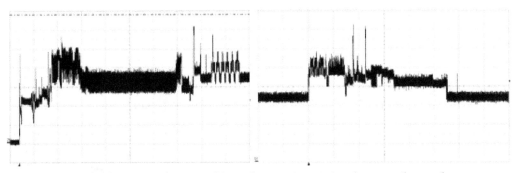

Fig. 2. At the left, the DSC current profile under turning-on in photography mode (100mA/div) is shown. Time base is set to 500ms/div. At the right, the DSC current profile under taking picture without flash (100mA/div) is shown. Time base is set at 500ms/div.

Fig.3. The peak current value is equal to 669mA and the transient time is equal to 430ms. At the right, the load current under the zoom function in movie mode is shown. The transient time is 630ms long and the current reaches a 691mA peak value. Only a few examples are shown. The load current profile of each available function is monitored and collected data are stored to be used in the DSC model. Critical functions for the hybrid power supplies are identified by the transient time and the current peak value. The worst-case function features the highest current peak value and the longest transient time. The power management system should prevent an excessive supercapacitor discharge under the worst-case.

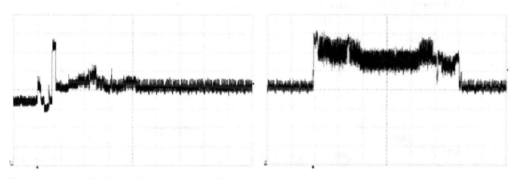

Fig. 3. At the left, the DSC current profile under the transition from photography to movie mode (100mA/div) is shown. Time base is set to 100ms/div. At the right, the DSC current profile under zoom function in movie mode (100mA/div) is shown. Time base is set at 100ms/div.

### 3.2 Measurements on the fuel cell

A measurement-based fuel cell model is proposed. The model matches the steady-state and dynamic behaviour of the stack under test, including temperature effects. The stack is accurately characterized to achieve high-accuracy. Even if a 10W, 9V Proton Exchange Membrane (PEM) fuel cell stack is tested, the model can be efficiently applied to each fuel cell type and power level. Steady-state measurements are performed. Iso-thermal steady-state V-I curves are measured at several temperature values ranging from 20°C to 40°C by 5°C step. Experimental data are collected and averaged over multiple measures. Average iso-thermal steady-state curves are plotted in Fig.4. A current limit is shown for each steady-state curve. If the fuel cell current overcomes the limit a flooding event occurs: hydrogen is not enough to produce the required current. Hence, water is sucked into the stack obstructing the membrane pores. A flooding event leads to an instantaneous voltage drop and the device is immediately turned-off. The limit is a function of the operating temperature since the reaction speed increase with increasing temperature. The fuel cell current is directly controlled by the power management system avoiding the flooding risk. The frequency response analysis is performed to model the dynamic behaviour of the stack under test. The impedance function is measured under linear conditions. A sinusoidal current is forced and the magnitude of the sinusoidal voltage is measured. The stack non-linearity is neglected while measuring the transfer function. Non-linear effects of operating temperature are neglected by performing measurements at a fixed temperature value of 30°C. Further, the steady-state curve non-linearity is neglected by a small-signal approximation. The average current value is fixed at 600mA and a sinusoidal signal of 200mAp-p is forced. In the fixed current region, at 30°C, the non-linearity of the steady-stare curve could be neglected. The goal is to model the impedance transfer function by a dynamic admittance which is connected in parallel with the steady-state model. The equivalent resistance of the steady-state model is then connected in parallel to the dynamic impedance. If the operating condition changes, the equivalent resistance of the steady-state model changes and the model is able to fit the dynamic behaviour of the stack under test. Experimental results validate the modelling technique. For further details, please refer to (Boscaino et al., 2009b). The Bode plot of the impedance magnitude is shown in Fig.5. On y-axis a linear scale is used. Since the system square wave response analysis shows no time constant lower than 1μsec, the magnitude of the transfer function is assumed constant for high frequency values.

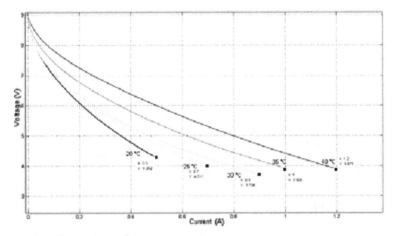

Fig. 4. Average iso-thermal steady-state curves.

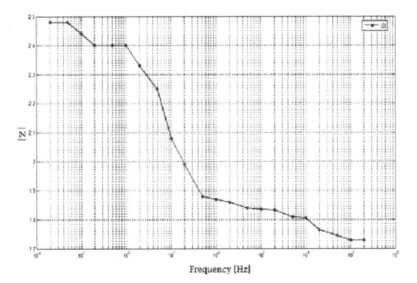

Fig. 5. Bode plot of the impedance magnitude.

Measurements are performed up to 200kHz. Temperature effects on the fuel cell performance are included in the fuel cell model. Further, the model reproduces the stack temperature as a function of the fuel cell current. The thermal response of the fuel cell is monitored as shown in Fig.6. Thermal measurements are performed by monitoring the current profile and the corresponding temperature profile.

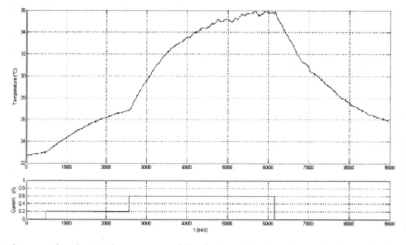

Fig. 6. At the top, the thermal response of the fuel cell stack. At the bottom, the corresponding current waveform.

## 4. The simulink model

A fuel cell – supercapacitor hybrid power supply is designed. The active hybrid source is modelled by a PSIM schematic as well as the power management control system. The primary source and the power load are modelled in Simulink environment. The top level model is shown in Fig.7. The *SimCoupler* block simulates the power management system implemented by a PSIM schematic which is fed by the fuel cell voltage and the load current signals and generates the output voltage of the hybrid supply.

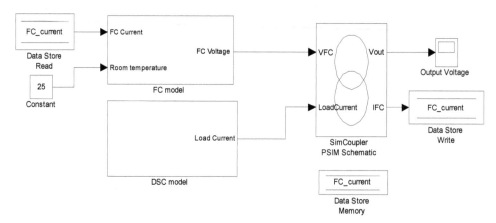

Fig. 7. Hybrid power supply model.

The fuel cell current is sensed within the PSIM schematic. The fuel cell current signal feeds the fuel cell subsystem by the means of data stores. In order to enable the signal analysis in Simulink environment too, the output voltage is considered as a Simulink Output Node.

## 4.1 The fuel cell model

The fuel cell model is shown in Fig.8. The stack voltage is generated as a function of the fuel cell current. The room temperature is introduced as an external parameter. The steady-state and dynamic sections are highlighted. The steady-state behaviour, including temperature effects, is modelled by the *Steady-state Section*. The dynamic section models the transient response by the means of an admittance which is connected in parallel to the steady-state model. Hence, the parallel connection is modelled by feeding the *Dynamic Section* by the output voltage and introducing a current node at the input of the *Steady-State Section*. By the means of data stores, the fuel cell voltage is stored as an output of the *Steady-State Section* and then read by the *Dynamic Section* as an input signal. The *Steady-State Current* is obtained by adding the *Dynamic Current* to the external fuel cell current.

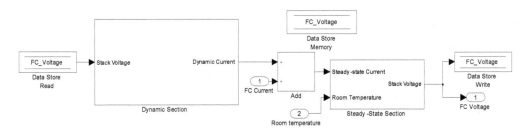

Fig. 8. The fuel cell model.

The steady-state section is modelled as shown in Fig.9. Note that the stack temperature is quite different from the room temperature. The room temperature is an external signal while the stack temperature is reproduced as a function of the fuel cell current by the temperature subsystem. The steady-state subsystem reproduces the stack voltage as a function of the fuel cell current $i$ and the reconstructed temperature value $T$. The steady-state subsystem is modelled as shown in Fig.10. The steady-state voltage at a temperature $T$ under a current $i$ is given by:

$$V(i,T)=V(i,Tref)+c(i,T-Tref) \tag{1}$$

where $V(i,Tref)$ is the steady-state voltage at a nominal temperature of 30°C under a current $i$ and $c(i,T-Tref)$ is the voltage correction term accounting for a temperature $T$ which is different from the nominal one.

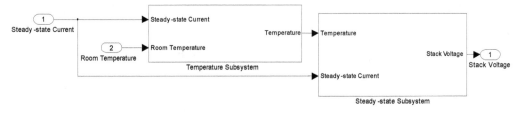

Fig. 9. The steady-state section model.

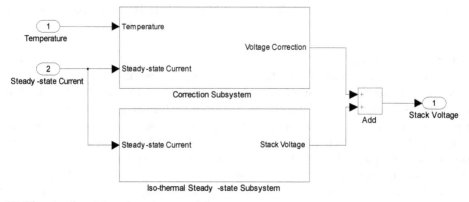

Fig. 10. The steady-state subsystem model.

The term *V(i,Tref)* is the output signal of the so-called I*so-thermal steady-state subsystem* while the correction term is generated by the *Correction subsystem*. No matter how many and which the involved physical, geometrical and chemical parameters are, the iso-thermal steady-state curve at the reference temperature is fitted by the law:

$$V(i, Tref) = A - B \cdot ln\left(1 + \frac{i}{c}\right) - D \cdot e^{\frac{i}{E}} \tag{2}$$

where A,B,C,D,E are the iso-thermal steady-state model parameters. Model parameters are obtained by fitting the experimental data with the aid of the *Curve Fitting Tool* which is available in MATLAB environment. The *Iso-thermal steady-state subsystem* is shown in Fig.11. The subsystem is fed by the Steady-state current and reproduces the stack voltage at the reference temperature of 30°C. The A parameter is modelled by the *constant* block. Exponential and logarithmic terms are generated by the subsystems shown in Fig.12. According to (2), parameters B, C, D and E are modelled by proper *Gain* blocks which are introduced in the signal chain. The gain value is entered by the parameter box. Exponential and logarithmic functions are implemented by *Math Function* blocks which are readily available in the Simulink library browser.

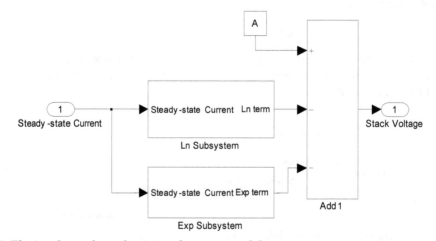

Fig. 11. The iso-thermal steady-state subsystem model.

The type of function is selected by the parameters box. The shift by 1 of the logarithmic argument is modelled by the *Bias* block. The shift constant is introduced by the block parameter box. Very complex functions, such as the logarithmic or exponential function are readily implemented by elementary blocks.

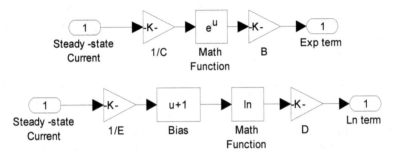

Fig. 12. The exponential and logarithmic subsystems.

The correction subsystem accounts for temperature effects. The subsystem is fed by the stack temperature and the fuel cell steady-state current, as shown in Fig.13.

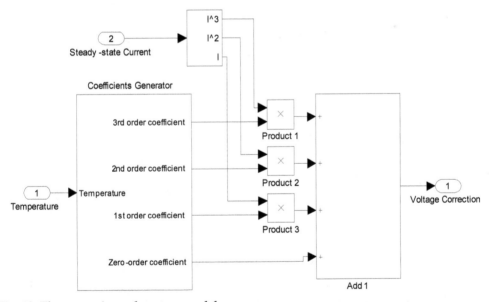

Fig. 13. The correction subsystem model.

According to (1) a set of average correction curves is obtained. If a point of the iso-thermal correction curve at the temperature $T$ is selected, the x-axis coordinate equals the fuel cell current $i$ and the y-axis coordinate the correction term $c(i, T\text{-}T_{ref})$. The correction subsystem is modelled by a bi-dimensional polynomial fitting of the average correction curves. A third-order polynomial is chosen for current-fitting while a fourth-order polynomial is chosen for temperature fitting. The correction curve at a temperature $T_1$ is fitted as a function of current by the polynomial law:

$$C(i, T_1) = C_0(T_1) \cdot i^0 + C_1(T_1) \cdot i^1 + C_2(T_1) \cdot i^2 + C_3(T_1) \cdot i^3 \qquad (3)$$

where C0, C1, C2 and $C_3$ are the polynomial coefficients. Each coefficient of the current polynomial is given by evaluating a fourth-order polynomial at the fixed input temperature. Coefficients at a fixed temperature $T_1$ are given by:

$$C_0 = C_{0,0} + C_{0,1}(T_1 - T_{ref}) + C_{0,2}(T_1 - T_{ref})^2 + C_{0,3}(T_1 - T_{ref})^3 + C_{0,4}(T_1 - T_{ref})^4$$
$$C_1 = C_{1,0} + C_{1,1}(T_1 - T_{ref}) + C_{1,2}(T_1 - T_{ref})^2 + C_{1,3}(T_1 - T_{ref})^3 + C_{1,4}(T_1 - T_{ref})^4$$
$$C_2 = C_{2,0} + C_{2,1}(T_1 - T_{ref}) + C_{2,2}(T_1 - T_{ref})^2 + C_{2,3}(T_1 - T_{ref})^3 + C_{2,4}(T_1 - T_{ref})^4 \qquad (4)$$
$$C_3 = C_{3,0} + C_{3,1}(T_1 - T_{ref}) + C_{3,2}(T_1 - T_{ref})^2 + C_{3,3}(T_1 - T_{ref})^3 + C_{3,4}(T_1 - T_{ref})^4$$

where Ci,j is the j-th order coefficient of the temperature polynomial which gives the Ci coefficient of the current polynomial. Fig.14 shows the iso-thermal correction curves after the fitting operation as a function of the fuel cell current.

The *Coefficients generator subsystem* is fed by the temperature signal and generates the coefficients of the current polynomial $C_0$, C1, C2 and $C_3$ corresponding to the input temperature value. Then the current polynomial is evaluated at the input current value. The subsystem is modelled by elementary blocks of Simulink libraries.

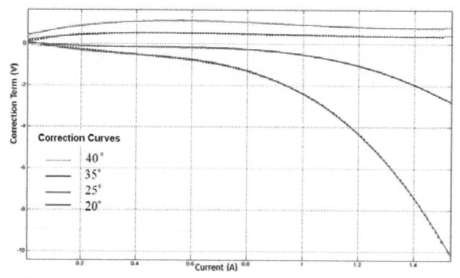

Fig. 14. Iso-thermal correction curves after current fitting procedure.

The model of the *Coefficients Generator* subsystem is shown in Fig.15. *Polyval* blocks are introduced to model a polynomial. Coefficients are set by the parameter box.

Hence, the current polynomial coefficients are evaluated by the *Coefficients Generator* subsystem which is fed by the temperature value *T*. The coefficients are applied to evaluate the current polynomial at the input current value *i*.

The thermal section is modelled by fitting the fuel cell thermal transient response by the *Parameters Estimation Toolbox* which is available in MATLAB/Simulink environment. The temperature profile is reproduced by the Temperature subsystem as a function of the fuel cell current. The thermal response of the fuel cell is modelled as:

$$a\frac{dT}{dt} = -(T - T_{room}) + g \cdot i + d \cdot i^2 \qquad (5)$$

where $T$ is the absolute stack temperature, $T_{room}$ the ambient temperature and $i$ the total current.

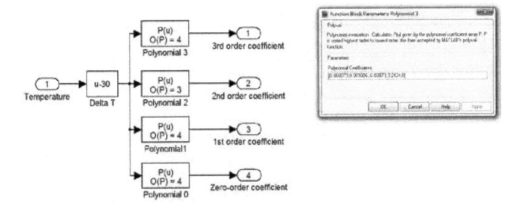

Fig. 15. The Coefficients Generator subsystem model.

The term $d\,i^2$ involves Joule heating to temperature $T$ and the term $g\,i$ models the evolved heat, the effects of the reaction rate and lost heat. Effects of room temperature are encountered by the first term in (5). Thermal model parameters $a$, $g$, $d$ are obtained by performed measurements with the aid of the *Parameter Estimation Toolbox*, which is available in Simulink environment. By Laplace transform, the (5) is solved as:

$$T(t) = T_0 \cdot e^{-\frac{t}{a}} + \frac{1}{a} \cdot e^{-\frac{t}{a}} \cdot \int_0^t T_{room}(t) \cdot e^{\frac{\tau}{a}} \cdot d\tau + \frac{g}{a} \cdot e^{-\frac{t}{a}} \cdot \int_0^t i(\tau) \cdot e^{\frac{\tau}{a}} \cdot d\tau +$$

$$+ \frac{d}{a} \cdot e^{-\frac{t}{a}} \cdot \int_0^t i^2(\tau) \cdot e^{\frac{\tau}{a}} \cdot d\tau \qquad (6)$$

In Fig.16 the temperature subsystem model is shown. The temperature signal is generated by adding integral and exponential terms. Each elementary term of (6) is modelled by Simulink library blocks. The *Clock* block generates the simulation time $t$. Exponential terms are evaluated by the means of the time $t$ and the *Math function* block. The integral operator is modelled by the *Integrator* block.

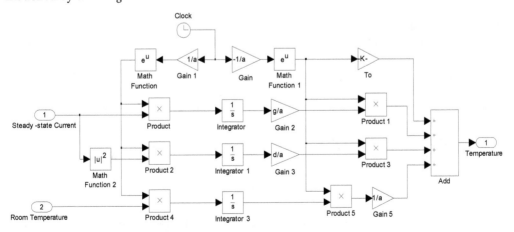

Fig. 16. The temperature subsystem.

According to (5), the thermal response to a current step is given by:

$$T(t) = T_0 + \left(T_{room} - T_0 + g \cdot I + d \cdot I^2\right)\left(1 - e^{-\frac{t}{a}}\right) \tag{7}$$

where I is the current step final value, $T_{room}$ is the room temperature and $T_0$ the initial temperature condition. A current step towards zero current leads to a thermal transient whose final value is the environment temperature $T_{room}$. According to (5), the thermal response to a current step towards zero current value is given by:

$$T(t) = T_0 + \left(T_{room} - T_0\right)\left(1 - e^{-\frac{t}{a}}\right) \tag{8}$$

By fitting the thermal response to a current step towards zero value, the $a$ parameter is obtained. The thermal response to the rising and falling-edge of a current step is monitored, obtaining $g$ and $d$ parameters.

As discussed, the dynamic subsystem models the impedance which is dynamically connected in parallel to the steady-state model. The approximated impedance transfer function is modelled by the *Transfer Function* library block, as shown in Fig.17. The dynamic voltage is obtained by subtracting the constant term of the steady-state model to the stack voltage. The dynamic current is generated by the transfer function block Y(s). The analysis of the model accuracy yields a maximum voltage error value of 0.2V, comparable to experimental data scattering. The thermal model error is limited to 1°C. Authors suggest (Boscaino et al., 2009b) for further details.

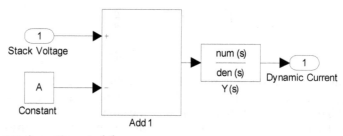

Fig. 17. The dynamic section model.

## 4.2 The DSC model

The load emulator is modelled in Simulink environment. Thanks to the powerful *Stateflow* toolbox a user-interactive model is implemented. During behavioural simulation, the designer is able to select a DSC function emulating the effective load device. The effective power consumption pattern corresponding to the actual user-selected function is generated by the load model. The designer is able to reproduce a random usage of the electronic load device.

No assumptions are made about the load current profile during behavioural simulations. System performances are already evaluated by computer simulations saving hydrogen reserve for preliminary tests and avoiding the risk of experimental fault conditions. The hybrid source life-cycle and fuel consumption could be tested as well as system stability and dynamic response.

The load device is modelled by a state machine: user-selections are modelled as events and the available functions are modelled as the finite states. The state-machine is implemented by the *Stateflow* toolbox which is available in Simulink library browser. The model is based on a Fujifilm S5500 digital still camera. Main sections of the DSC model are highlighted in Fig.18: *User Interface, Mux* and *DSC subsystem*. Manual switches are included in the *User Interface* section. Each manual switch is labelled by a specific DSC function. By acting on a manual switch, the designer generates events for the state-machine thus running the corresponding function. Scalar events are collected by the *Mux* subsystem into *DSC subsystem* input vector. In the *DSC subsystem*, the state-machine and the output section are modelled.

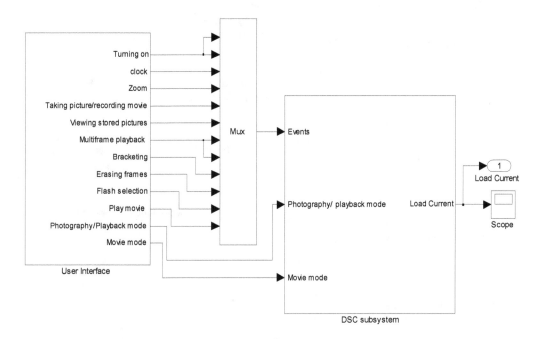

Fig. 18. MATLAB/Simulink DSC model.

The state-machine receives events and conditions. The event immediately triggers a state transition. Under the clock rising-edge conditions are verified thus enabling a state transition. Events and conditions are generated in the User-Interface subsystem as shown in Fig.19.

Manual switches are collected within the subsystems. Take a look to the *Conditions* subsystem shown in Fig.20. Photography or playback mode is selected by double-clicking on the corresponding manual switch. A condition for the state-machine is generated by the manual switch. If the condition is true, corresponding to the logic value 1, the state-machine is driven toward the Photography Mode. Otherwise, the system is driven towards the Playback Mode. Even though events are generated instead of conditions, the Events subsystem features a similar architecture, as shown in Fig.21. End-user actions on the digital camera selector switch are marked as conditions while actions on buttons are marked as events. Each manual switch within the Events subsystems emulates a physical button of the digital camera under test. As shown Fig.22, the *DSC subsystem* is modelled by a state diagram and an output section.

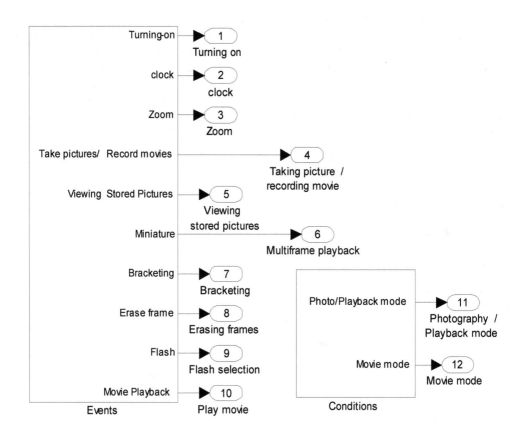

Fig. 19. The user-interface subsystem.

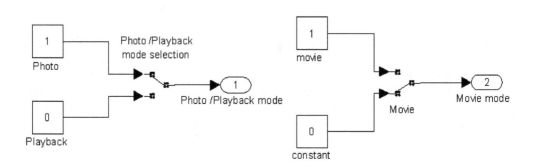

Fig. 20. Conditions.

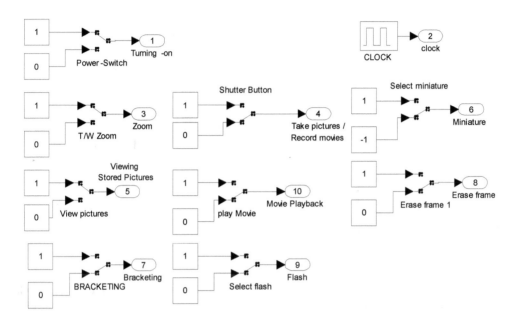

Fig. 21. Events.

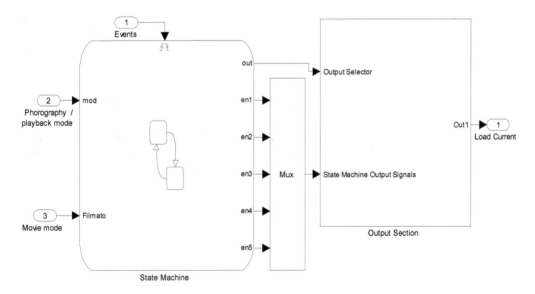

Fig. 22. The DSC subsystem.

A set of look-up tables (LUT) is included in the output section, as shown in Fig.23. Each LUT stores the current profile of a DSC function. The state diagram generates five control signals which are multiplexed in a control vector by the *Mux* block. In the output section, a decoder is fed by the control vector and generates the enable signals for LUTs. Only one enable signal at once is active. If a function is selected by user, the corresponding state in the state diagram is active and then the corresponding LUT in the output section is enabled and sequentially read. The output bus is connected to the active LUT output by the multiport switch which is driven by the output selector signal.

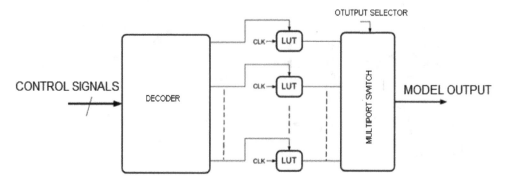

Fig. 23. The output section.

The state machine is modelled by the Stateflow toolbox. Only one state at once is enabled. Three standby modes of operation are available: playback, photography and movie mode. Each one could be maintained indefinitely active even if no function is selected by user. Each mode allows the user to select a limited number of functions. For example, under photography mode taking picture, bracketing, taking picture with flash are available while stored pictures could be viewed if the DSC is in playback mode only. Then, the state machine is intrinsically hierarchical. Fig.24 shows a schematic block diagram of the state machine. Under system start-up, the DSC is driven towards the OFF state by the default transition. No function is available in the OFF state and the output current is null. By double-clicking on the power switch button, the on_switch event is generated and the DSC is driven towards the ON state. The ON-state architecture is shown in Fig.25. Within the ON state, a decision point is set. The position of the mode selector enables playback or photography mode. Note that for the DSC under test, the movie mode is selected by acting on two manual selectors: the first is set on photography mode and the latter selects movie mode. The model reflects the operation of the DSC under test. Within each state, available functions are modelled as states. Authors suggest (Boscaino et al., 2008) for further details.

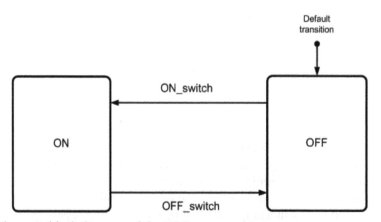

Fig. 24. A schematic block diagram of the DSC state machine.

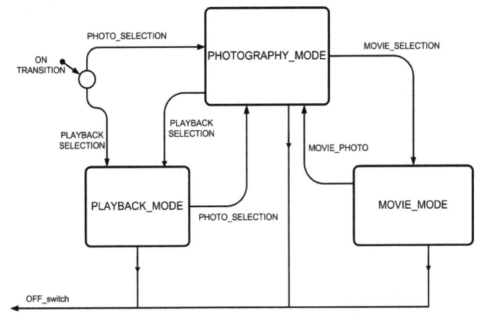

Fig. 25. The ON state architecture.

## 4.3 PSIM schematic

The active hybrid source and the power management system are modelled in PSIM environment. The stack current is directly controlled to improve dynamic performances. The fuel cell load is levelled by the power management system. The supercapacitor meets the load transient power demand, required for each function running while the fuel cell supplies the standby current of each available mode of the digital camera: playback, photography and movie mode. The equivalent fuel cell load is rather constant except for mode transitions of the DSC. Since the standby current required by the DSC depends on the selected mode of operation, a multiple steady-state control is applied. The fuel cell current is limited to the required stand-by current. Hence, the information about the actual DSC mode is required by the control system. The information is derived by a digital subsystem by monitoring the output voltage and comparing the last few samples.

Fig.26 shows the PSIM top level model. The DC-DC converter and supercapacitor bank are included in the *Hybrid Source* subsystem. The *Control System* is fed by the output voltage and the current sense signal. The output of the controller feeds the *Driver* subsystem which drives the synchronous buck MOSFETs. The Simulink model of the fuel cell reproduces the fuel cell voltage by sampling the fuel cell instantaneous current. In PSIM circuit, the fuel cell voltage is considered an input node from Simulink while the fuel cell current signal is an input signal for the Simulink model. Since Simulink input and output nodes behaviour as voltage sources, the fuel cell current is converted to a voltage signal by a current controlled voltage source which is included in the *Hybrid source* subsystem. The DSC model reproduces the current profile corresponding to the user-selected function. The load current signal is thus obtained by the *Load Current* Simulink input node. Since Simulink input nodes behaviour as voltage sources in PSIM schematics, the signal is converted to a current signal by the means of a voltage controlled current source.

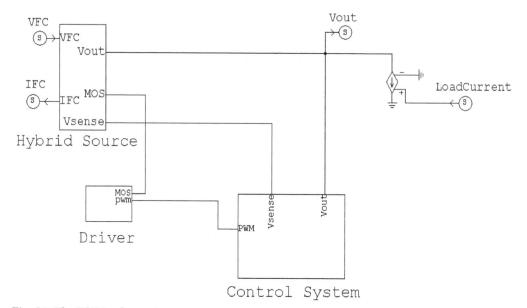

Fig. 26. The PSIM schematic.

Fig.27 shows the active hybrid source. The Fuel cell is connected in series with a buck converter and the supercapacitor bank. Even under the worst-case scenario, the supercapacitor should supply the load current under function transients avoiding any accidental turning-off of the electronic device. The supercapacitor value has been designed to ensure a negligible supercapacitor discharge under the worst-case obtaining a 4.6F value. The load nominal voltage is 6V while turning-off voltage is 4.2V. Since the maximum value of the supercapacitor voltage is 2.5V, three supercapacitors are connected in series to match the load nominal voltage. Two parallel branch of Maxwell PC5 and PC10 supercapacitors are connected. A charge balancing circuit is provided on the experimental prototype. Low ESR capacitors are necessary to limit the output voltage ripple avoiding high inductance values. A sense resistor is introduced to sense the instantaneous inductor current waveform generating the $V_{sense}$ signal.

The control loop architecture is shown in Fig.28. The output voltage is controlled by a Pulse Width Modulation (PWM) control loop. The load levelling function is implemented by a current loop. If a DSC function is selected, the current loop limits the fuel cell current at the standby current value of the actual DSC mode of operation. For example, if bracketing is selected the DSC is in photography mode. The buck inductor current and hence the fuel cell current is limited at the photography standby current value by the current control loop. The pulsed current is supplied by the supercapacitor bank over the entire function runtime. Then, under function running the current loop is active and the current loop action must prevail over the voltage one. Further, the control loop should detect the mode of operation of the power load. The voltage reference subsystem detects the actual DSC mode generating the corresponding voltage reference of the current loop.

Error amplifiers are implemented as Transconductance Operational Amplifiers (OTA). The OTAs are modelled by elementary blocks of PSIM libraries. The differential input is modelled by an adder, the OTA gain is modelled by a multiplier and the linearity range is modelled by a signal limiter.

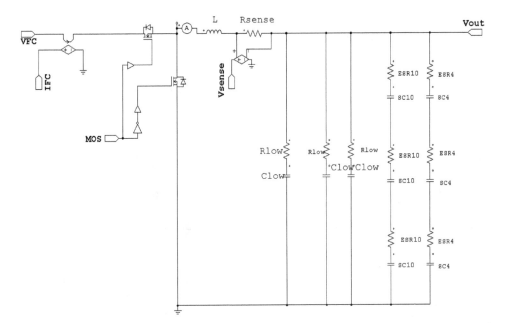

Fig. 27. The hybrid source model.

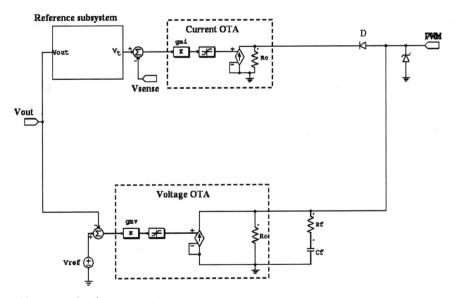

Fig. 28. The control subsystem.

The output current signal is obtained by a voltage controlled current source connected in parallel to the output resistance of the OTA. The compensation action is implemented by the RfCf network. Under steady-state condition, the inductor current is below the current limit and hence the cathode node of the diode D is tied to high-level. The diode D is turned-off. Under load transients, the inductor current tries to overcome the current limit, the cathode voltage is tied to lower level thus turning-on the diode D. By a proper design of the control loop parameters, if the diode is turned-on the current loop action prevails over the voltage one. The interface diode turns on under the condition:

$$I_I \cdot R_0 \leq R_0 \cdot I_V - V_\gamma \tag{9}$$

where $V_\gamma$ is the diode threshold voltage, $I_I$ is the current OTA output signal, $I_V$ is the voltage OTA output signal. The turn on condition can be expressed as:

$$g_{m,I} \cdot e_I \cdot R_0 \leq g_{m,V} \cdot e_V \cdot R_0 - V_\gamma \tag{10}$$

where $e_V$ and $e_I$ are the voltage and current differential input signal, respectively, while $g_{m,I}$ and $g_{m,V}$ are the current and voltage OTA gains, respectively.

The current loop action should prevail over the voltage loop action even if the voltage loop is completely upset so ensuring the load levelling function. Hence, assuming the total upset of the voltage loop, the worst-case diode turning-on condition is given by:

$$g_{m,I} \cdot e_I \cdot R_0 \leq R_0 \cdot I_{LIM} - V_\gamma \tag{11}$$

where $I_{LIM}$ is the voltage loop OTA maximum output current and $e_I$ is the desired tolerance on current control. The tolerance is fixed at one half of the steady-state inductor current ripple to achieve a peak current control. The OTA gain of the current loop is designed to verify the (11). Under a load transient, the load current $I_{LOAD}$ exceeds the reference value $I_{L,th}$. Hence, the inductor current is limited to $I_{L,th}$ and the current $I=I_{LOAD}-I_{L,th}$ is supplied by the supercapacitor until the end of transient. Consequently, under a load transient the fuel cell current is maintained constant for hybrid power source optimization: the fuel cell supplies a constant load except for reference value changes. The OTA gain of the voltage loop is chosen by frequency domain analysis to ensure system stability and an adequate bandwidth.

The operating mode of the power load is detected by the reference subsystem by sampling the output voltage only. The reference subsystem is able to detect mode transitions neglecting function transients. Under function transients, the reference voltage is kept constant while under mode transients the voltage reference is adapted to the new mode of operation. Unlike mode transients, function transients feature a limited run time. The subsystem clock period is longer than the highest runtime and hence the control loop is insensitive to function transients.

Fig.29 shows the reference subsystem model which is implemented by logic gates. The output voltage sample is compared with two threshold levels, which are accurately designed to prevent an excessive discharge. Information about the instantaneous charge state of the supercapacitor bank is obtained. The output voltage sample is coded by the logic signals up, mid and down: up is active if the output voltage is over both levels, mid if the sample is in the middle, down if the output voltage is lower than the lowest level. By comparing the two last samples of binary signals, the charge process is monitored and the DSC mode is detected.

Four possible steady-states, coded by s1 and s2 bits, are provided: 240mA, corresponding to DSC playback mode standby power consumption, 390mA, corresponding to DSC photography mode standby power consumption, 480mA, corresponding to DSC movie mode standby power consumption, 700mA, corresponding to a safety state. If an excessive supercapacitor discharge occurs, the safety state will prevent any accidental DSC turning-off. Four possible actions, coded by $a$ and $b$ bits, are provided: increase by one, increase by two, decrease by one and keep constant the reference voltage. The subsystem decides which action should be performed on the voltage reference. The new voltage reference bits are determined by the old reference bits and by the action bits. The reference bits feed a digital-to-analog converter which outputs the analog voltage reference value. The proposed algorithm reproduces information about the DSC state by monitoring the charge and discharge profile of the supercapacitor bank. The PWM control technique is implemented within the *Driver* subsystem shown in Fig.30. The error signal is compared with the external saw-tooth voltage $V_{ramp}$ and the PWM signal is generated. The PWM signal and the clock signal feed the D flip-flop with asynchronous reset input which generates the driver signal for both MOSFET gates of the buck converter. Authors suggest (Boscaino et al., 2009a) for further details.

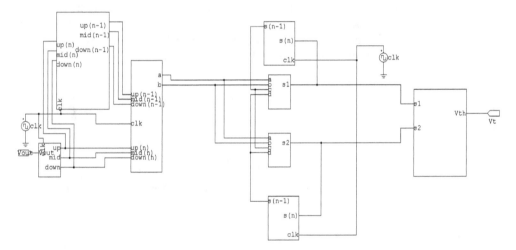

Fig. 29. The reference subsystem.

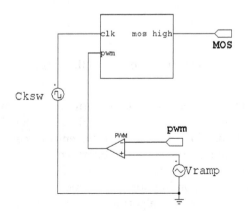

Fig. 30. The driver subsystem.

## 5. Simulation and experimental results

The efficiency and accuracy of the proposed simulation set-up is confirmed by the comparison between simulation and experimental results. Simulation results accurately match experimental ones. The output voltage profile depends not only on the selected function but above all on the "history" of the power supply system since the turning-on. Yet, the system is tested since the simulation step with the effective load consumption profile. Long-time simulation tests are performed to show the robustness of the power management algorithms. The system prevents any accidental turning-off of the digital still camera improving the hybrid source performances. The hybrid source features the same dynamic response of the supercapacitor bank under pulsed current profile, the life cycle is closely related to the hydrogen reserve and quite independent of the supercapacitor state of charge. After all, the hybrid source features the high power capability of the supercapacitor bank and the high energy capability of the fuel cell. The performances of the hybrid source are not affected by drawbacks of both primary sources. The inductor current and the output voltage waveforms under no-load conditions are shown in Fig.31. The output voltage is controlled by the voltage loop while the current loop is not active.

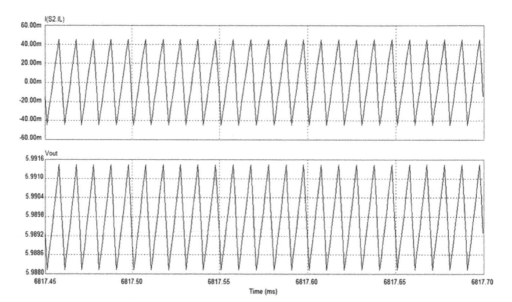

Fig. 31. At the top the inductor current, at the bottom the output voltage waveforms under no-load conditions.

The inductor current and the interface diode current waveforms under current loop limitation are shown in Fig.32 the current limit is fixed at 240mA, corresponding to the playback mode of operation. When the inductor current overcomes the 210mA limit, the interface diode is forced into conduction. The tolerance on the current control is 30mA within the tolerance window of 40mA designed by (11).

Simulation results under turning-on in photography mode are shown to better understand the efficiency of the designed control algorithm. Turning-on in photography mode is a critical transient since the transient time is quite long and peak current value is higher and higher than the average value. Figures from 33 to 36 show the system behaviour under the turning-on function. Take a look to the evolution of the system from start-up till the end of transient. At the top, the pulsed current load and the inductor current, at the bottom the output voltage waveforms are shown. Under start-up conditions, the system starts at the lowest threshold value corresponding to the playback mode (240mA). Yet, the digital camera is turned-on in photography mode corresponding to a higher current limit (390mA). The fuel cell supplies the standby current of the playback mode while the pulsed current is supplied by the supercapacitor bank, as shown in Fig.33.The turning-on function in photography mode lasts 4.6s. Until the effective mode of operation is detected, the inductor current is limited to the playback mode current threshold, as shown in Fig.33. In the intermediate part of the function transient the main peak of the load current is entirely supplied by the supercapacitor, as shown in Fig.34. The supercapacitor bank is constantly discharged. Even under stress, the output voltage does not drop below the turning-off limit, which is equal to 4.2V, thanks to an accurate design of the power stage. Fig.35 shows the system behaviour under the last part of the turning on function, corresponding to the lens movement. The inductor current is already limited at the playback mode current limit. At the end of turning-on in photography mode, the actual mode of operation is detected by the reference subsystem as shown by Fig.36. The inductor current limit is increased up to the photography mode standby current value. No function is actually selected by the end-user. The standby current of the photography mode is supplied by the fuel cell thus interrupting the supercapacitor discharge. The safety state is not reached since no excessive discharge has been detected by the reference subsystem.

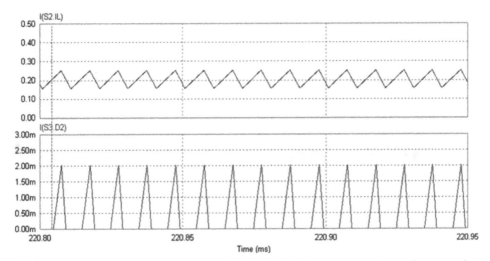

Fig. 32. At the top the inductor current, at the bottom the diode current waveforms under current limitation.

Experimental results under turning-on in photography mode are shown in Fig.37. The output voltage (C1, 1V/div), the load current (C2, 200mA/div) and the inductor current (C3, 200mA/div) are shown. Time base is fixed at 2s/div. As shown by simulation results, the system does not detect the actual mode of operation until the end of transient. Hence, the transient current is supplied by the supercapacitor bank while the standby current is supplied by the fuel cell. Then, the actual mode is detected and the current limit is increased.

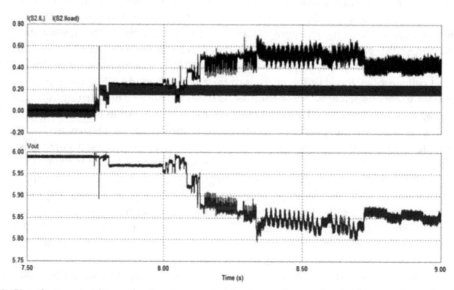

Fig. 33. Simulation results under turning-on in photography mode. At the top the pulsed load current and the inductor current, at the bottom the output voltage waveforms at the beginning of function transient are shown. The system starts at playback mode current limit and the actual state is not yet detected.

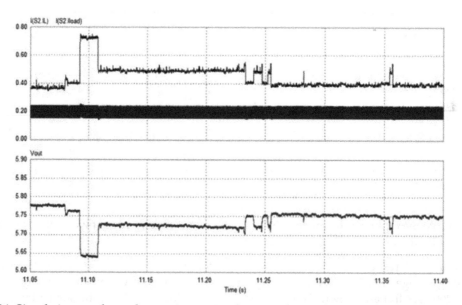

Fig. 34. Simulation results under turning-on in photography mode, intermediate part of function transient. At the top the pulsed load current and the inductor current, at the bottom the output voltage waveforms during the peak power consumption are shown. The DSC state is not yet detected.

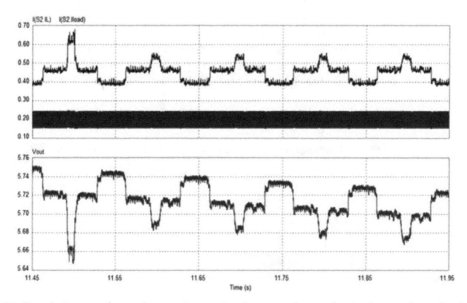

Fig. 35. Simulation results under turning-on in photography mode. At the top the pulsed load current and the inductor current, at the bottom the output voltage waveforms during at the end of function transient are shown. The actual state is not yet detected.

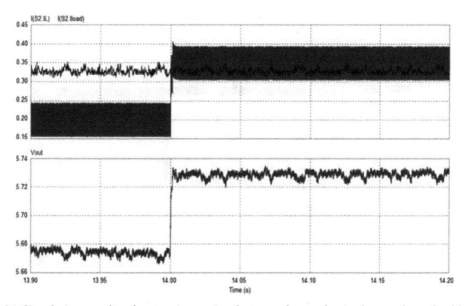

Fig. 36. Simulation results after turning on in photography mode. At the top the pulsed load current and the inductor current, at the bottom the output voltage waveforms are shown. The DSC state is detected and the current threshold is increased to allow the supercapacitor recharge.

Note that the instant at which the reference subsystem detects the actual DSC mode of operation is not related to the transient itself but above all on the history of the system since the start-up condition. Experimental results under turning-on in photography mode in the worst-case are shown in Fig.37. In this case, the actual mode of operation is detected after the end of transient. In Fig. 38, the actual mode of operation is detected before the end of transient. The discharge of the supercapacitor is thus limited to the beginning of the function transient.

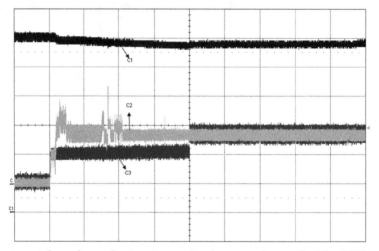

Fig. 37. Experimental results under turning on in photography mode. The DSC state is detected at the end of function transient. On C1 the output voltage (1V/div), on C2 the load current (200mA/div), on C3 the inductor current (200mA/div) are shown. Time base is set at 2s/div.

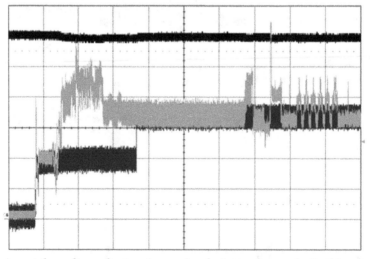

Fig. 38. Experimental results under turning on in photography mode. In this case the actual mode is detected before the end of transient. On C1 the output voltage (1V/div), on C2 the load current (200mA/div), on C3 the inductor current (200mA/div) are shown. Time base is set at 2s/div.

In Fig.39 simulation results under bracketing function are shown. At the top, the pulsed load current and the buck inductor current are shown. At the bottom, the output voltage is shown. The inductor current which is a scaled version of the fuel cell current is limited at the photography stand-by current limit over the entire transient time. The fuel cell supplies the standby current of the actual DSC mode of operation while the supercapacitor bank supplies the instantaneous peak power. The output voltage is affected by the supercapacitor discharge. Yet, any accidental turning-off is avoided thanks to the accurate design of the power stage. Further, if no-function is selected by user the supercapacitor bank will be recharged driving back the system to the nominal working conditions.

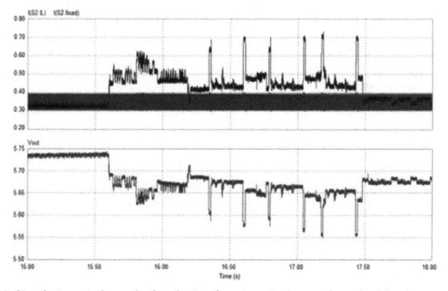

Fig. 39. Simulation results under bracketing function. At the top the pulsed load current and the inductor current, at the bottom the output voltage waveforms are shown.

Fig.40 shows simulation results under bracketing function. In this case, the system has not yet detected the actual mode of operation of the DSC. At the top the pulsed load current and the inductor current, at the bottom the output voltage waveforms are shown. The system is still in playback mode and the inductor current is limited to 240mA, lower and lower than the actual standby current. This is the worst-case for bracketing function. The fuel cell supplies the playback mode current while the supercapacitor is heavily discharged supplying both the pulsed current and the difference between the photography and playback standby currents. Yet, the discharge is not yet critical for the DSC working. Note that this condition is closely related to the history of the system.

Experimental results under bracketing function are shown in Fig.41. The output voltage (C1, 1V/div), the load current (C2, 200mA/div) and the inductor current (C4, 200mA/div) are shown. The time base is fixed at 500ms/div. Note the high accuracy of simulation results. The DSC current profile is accurately reproduced allowing the designer to test the control algorithms as closely as possible to the effective system behaviour. Since specific hybrid source features as life-cycle and transient response are directly controlled, high accuracy modelling is a key step to achieve high-efficiency power management control algorithms. As expected by the performed behavioural simulations, experimental results confirm the high efficiency of the implemented algorithms.

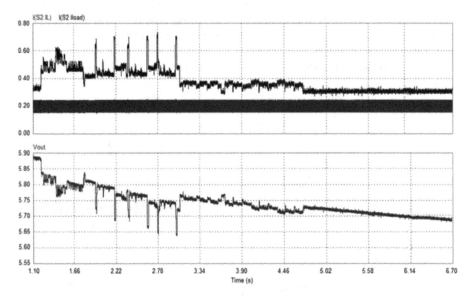

Fig. 40. Simulation results under bracketing function. The actual mode is not yet detected and the playback mode current limit is active. At the top the pulsed load current and the inductor current, at the bottom the output voltage waveforms are shown.

Simulation results under zoom function in playback mode are shown in Fig.42. At the top, the pulsed load current and the buck inductor current are shown. At the bottom, the output voltage is shown. The zoom function is not a critical function for the hybrid power supply. The inductor current is limited at the playback mode threshold. At the end of the transient time, the output voltage reaches again the nominal value. The net discharge of the supercapacitor bank could be neglected under the selected function running. The fuel cell supplies the standby current of the actual DSC mode of operation while the supercapacitor bank supplies the instantaneous peak power.

Fig.43 shows simulation results under multiple execution of zoom function in playback mode. At the top, the pulsed load current and the inductor current, at the bottom the output voltage waveforms are shown. In this case, the system has previously detected an excessive discharge of the supercapacitor bank. The current limit has been previously increased up to the movie mode threshold to allow a fast recharge. Yet, zoom function is actually selected by user and repeated several times. Note that in this case, the recharge is supplied by the fuel cell itself. The net recharge current is equal to the difference between the limited inductor current and the load current.

Experimental results under zoom function are shown in Fig.44. The output voltage (C1, 1V/div), the load current (C2, 200mA/div) and the inductor current (C3, 100mA/div) are shown. Time base is set to 20ms/div. As expected by simulation results, the output voltage is not affected by the zoom running.

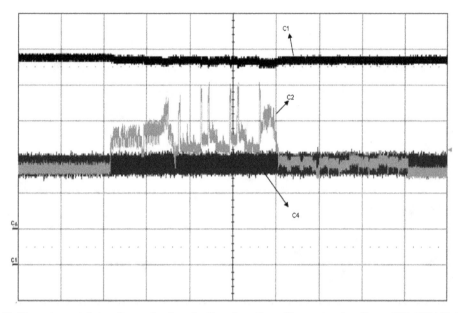
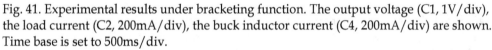

Fig. 41. Experimental results under bracketing function. The output voltage (C1, 1V/div), the load current (C2, 200mA/div), the buck inductor current (C4, 200mA/div) are shown. Time base is set to 500ms/div.

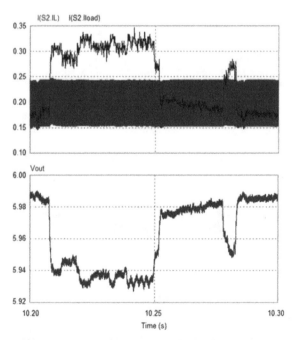

Fig. 42. Simulation results under zoom function in playback mode. At the top, the pulsed load current and the inductor current, at the bottom the output voltage waveforms are shown.

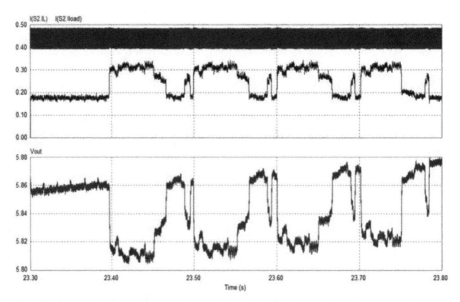

Fig. 43. Simulation results under zoom function in playback mode. The current limit is higher than the playback mode threshold to allow a fast recharge of the supercapacitor bank. At the top, the pulsed load current and the inductor current, at the bottom the output voltage waveforms are shown.

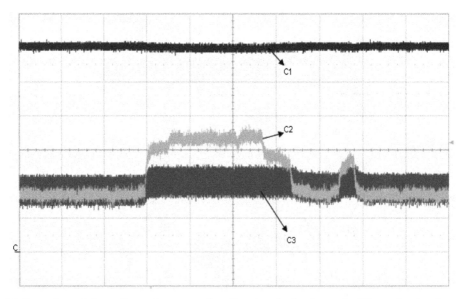

Fig. 44. Experimental results under zoom function in playback mode. The output voltage (C1, 1V/div), the load current (C2, 200mA/div), the buck inductor current (C3, 100mA/div) are shown. Time base is set to 20ms/div.

## 6. Conclusion

In this chapter, a high accuracy simulation set-up for hybrid power supplies is proposed. As an example, a fuel cell-supercapacitor hybrid power supply for a digital still camera is designed and tested. The system is modelled and tested in Simulink environment with the aid of *SimCoupler Module* for PSIM/Matlab co-simulation. The power management controller is implemented in PSIM environment while the fuel cell and the power load are accurately modelled by the means of Simulink library elements. A laboratory prototype has been realized and tested. Simulation results accurately match experimental results, highlighting the high efficiency of the proposed technique. The proposed modelling approach is suitable for each power supply system thanks to the powerful tools available in Matlab suite. Thanks to the powerful modelling approach, the need of a warm-up phase is highlighted since the simulation step. The fuel cell temperature has to be increased up to 40°C before the system start-up. Authors are working on an innovative power management system which overcomes this drawback by controlling the fuel cell temperature under start-up conditions.
The modelling approach is still a key design tool.

## 7. Acknowledgement

Authors would like to thank ST Microelectronics for supporting the research activity.

## 8. References

Alotto, P., Guarnieri, M. & Moro, F. (2008). Modelling and control of fuel cell-battery hybrid power systems for portable electronics. *Proceedings of IEEE Universities Power Engineering Conference,* ISBN: 978-1-4244-3294-3, Padova, Italy, December 2008.

Boscaino, V., Capponi G., Livreri, P. & Marino, F. (2008). Measurement-based load modelling for power supply system design, *Proceedings of IEEE International Conference on Control and modelling for Power Electronics,* ISBN: 978-1-4244-2550-1, Zürich, Switzerland, August 2008.

Boscaino, V., Capponi G. & Marino, F.(2009a). Experimental test on a fuel cell-supercapacitor hybrid power supply for a digital still camera, *Proceedings of IEEE Universities Power Engineering Conference,* ISBN: 978-1-4244-6823-2, Glasgow, Scotland, Sept. 2009.

Boscaino, V., Capponi, G. & Marino, F. (2009b). A steady-state and dynamic fuel cell model including temperature effects. *Proceedings of IEEE International Conference on Industrial Electronics,* ISBN: 978-1-4244-4648-3, Porto, Portugal, November 2009.

Gao L., & Dougal, R.A. (2003). Active Power Sharing in Hybrid Battery/Capacitor Power Sources. *Proceedings of Applied Power Electronics Conference and Exposition,* ISBN: 0-7803-7768-0, Miami, Florida, February 2003.

Lin, C.J., Chen, A.Y.T., Chiou, C.Y., Huang, C.H., Chiang, H.D. , Wang, J.C. & Fekih-Ahmed, L.(1993). Dynamic load models in power systems using the measurement approach. *IEEE Transactions on Power Systems,* vol. 8, no. 1, (February 2003), pp. 309–315, ISSN: 0885-8950.

# 3

# Calculating Radiation from Power Lines for Power Line Communications

Cornelis Jan Kikkert
*James Cook University*
*Queensland*
*Australia*

## 1. Introduction

The cost of electricity charged to consumers in Australia is predominantly set by state governments and does not vary depending on the actual cost that the electricity retailer has to pay the generators. This does not encourage users to reduce power at times of peak demand and the ratio between the peak demand and the average demand is growing. In the long term, user pays principle must apply and the cost of electricity charged will need to vary depending on the actual cost of production. This requires a communication system to be produced firstly to determine the date and time electricity is used, so that the usage and actual cost can be matched, and secondly to inform the consumer of the actual cost of electricity, so that appropriate actions can be taken, like turning off air conditioners when the cost per kW becomes high.

There are two ways that such a communication system can be developed. Firstly it can use the PLC for carrying the communication signals. Secondly the communication systems can be carried by third party suppliers. The WiMAX radio system currently being rolled out in Australia in Victoria and mobile phone based systems are examples of such systems. Using PLC has the advantage that the communication system is fully under the control of the electricity company and no fees need to be paid to third party operators.

In addition it is possible for the electricity supplier to use their power lines to provide BPL, giving internet access to customers, by putting communication signals up to 80 MHz onto the power lines. The frequency range used and bandwidth required for BPL is very different than that for PLC and the radiation losses are very different for BPL and PLC.

When a power line is used for transmitting communication signals, signal power is lost through electromagnetic radiation. This may cause interference to other users. There is a concern about the amount of radiation from power lines (ACMA 2005, 2007), with amateur radio operators claiming BPL creates excessive levels of radiation. Unfortunately much of the amateur radio literature does not distinguish between BPL and PLC. The second loss factor is resistive losses, which due to the skin depth causes an increasing attenuation with frequency.

The Matlab® program described in this paper and the results obtained from this program provide a rigorous basis for determining both the resistive losses and the radiation obtained from power lines. Electricity distribution companies and other communication providers can use this data to determine the feasibility of power line communications in their region.

SWER power lines are used in remote areas and are single phase lines where the ground is relied upon for the return path. As these power lines can be up to 300 km long, there is a significant economic incentive to read the power meter for those customers remotely since a 600 km trip on dirt roads is likely to cost more than the customer's electricity bill. To achieve this JCU has been carrying out a research project for the Queensland electricity retailer ERGON for developing a PLC communication system for SWER lines.

## 2. Fundamental principles

### 2.1 Line radiation theory

The existing travelling wave model of a long wire antenna (Balanis, 1997, Walter, 1970 and Ulaby, 1994) assumes that the antenna is less than 10 $\lambda$ long and the current along the line is constant, implying that no power is lost in radiation. Neither of these assumptions apply for power lines, so that a radiation model based on fundamental equations had to be developed.

The power radiated for an infinitesimal current element, also called an elementary doublet is described in most antenna books. The model presented in this paper cascades many of these infinitesimal current element models to make up the whole transmission line. The power radiated by a small length of conductor of length $l$ in the far field, is dependent on the current in the conductor, $I_0$, and is shown in (Balanis, 1997) to be:

$$P_{rad} = \eta \frac{\pi}{3} \left( \frac{l I_0}{\lambda} \right)^2 \tag{1}$$

where $\eta_0$ = 120 $\pi$, is the $\lambda$ wavelength. Antenna radiation patterns for long wire antenna Because the line is long, $I_0$ will not be constant. Three factors affect $I_0$:

1. $I_0$ is reduced by dissipation due to series resistance. This series resistance varies with frequency due to the skin depth varying with frequency.
2. $I_0$ is altered in phase, along the line, due to wave propagation.
3. $I_0$ is progressively reduced because power is radiated.

Using the conventional spherical coordinates r, $\theta$ and $\varphi$, the far-field electric field $E_\theta$ and the far-field magnetic field $H_\varphi$ for this elementary doublet is given by (Balanis 1997) as:

$$E_\theta = \frac{j\eta_0 k I_0 l e^{-jkr} \sin(\theta)}{4\pi r} = \frac{j\eta_0 I_0 l e^{-jkr} \sin(\theta)}{2\lambda r} = \frac{j60\pi I_0 l e^{-jkr} \sin(\theta)}{\lambda r} \tag{2}$$

$$H_\varphi = \frac{j k I_0 l e^{-jkr} \sin(\theta)}{4\pi r} = \frac{j I_0 l e^{-jkr} \sin(\theta)}{2\lambda r} \tag{3}$$

where the wave number, $k = 2\pi/\lambda$. Note that since the current $I_0$ has a magnitude and phase, both $E_\theta$ and $H_\varphi$ are complex quantities.

$$P_{rad} = \frac{1}{2} \text{Re} \left( \oiint_s E_\theta \times H_\varphi ds \right) \tag{4}$$

The far-field power radiated is obtained by integrating the Poynting vector over a closed surface, chosen to be a sphere of radius r, where r is much larger than the wavelength and the length of the power line dimensions as shown in equation (4). For an elementary doublet in free space, the inner integrand is independent of $\varphi$, resulting in a constant of $2\pi$ for that integration, so that:

$$P_{rad} = \frac{\pi}{2\eta_0} \int_0^{2\pi} |E(\theta)| r^2 \sin(\theta) d\theta = \frac{1}{240} \int_0^{2\pi} |E(\theta)| r^2 \sin(\theta) d\theta \qquad (5)$$

For a very long wire antenna, such as a power line, the magnitude and phase of the current $I_0$ and $E_\theta$ at a position along the line is iteratively determined by considering the radiation and resistive line losses up to that point. This is then be used to determine $E_\theta$ caused by the elementary doublet at that point. At a point along the line, the $E_\theta$ vector from the elementary doublet is added to the cumulative $E_\theta$ vector, to produce the new cumulative $E_\theta$ vector. For a line in free space, the cumulative fields are independent of $\varphi$, so that (5) can be used to evaluate the radiated power.

If the conductor is not in free space, but is a single power line, a distance h above an ideal groundplane, then equal and opposite image current will be induced below the groundplane. That changes the radiation pattern, so that the radiation is no longer independent of $\varphi$. A Medium Voltage (11 kV to 33 kV) single power line is typically 7 m above the ground. If this is used for power line communication signals up to 1 MHz, then the power line is far less than a quarter wavelength above the ground. The far-field electric field for an elementary doublet a distance h above a perfectly conducting ground plane will be:

$$E(\theta,\varphi) = E_\theta(2jSin(\frac{\pi 2h}{\lambda}Cos(\varphi)) = -\frac{120\pi I_0 le^{-jkr} \sin(\theta)}{\lambda r} Sin(\frac{\pi 2h}{\lambda}Cos(\varphi)) \qquad (6)$$

The radiation pattern from the elementary doublet will then have a null along the conductor ($\theta=0$), as it had for the free space conductor and in addition it will have a null along the groundplane ($\varphi=90°$) and a maximum perpendicular to the groundplane ($\theta = \varphi = 90°$). The field in this direction depends on 2h, the separation between the line and the return path. For SWER lines, the ground plane is not a perfectly conducting plane and the distance of the ground return path below the ground level is given by Carson's equation (Carson, 1926, Wang & Liu, 2001, Deri & Tevan, 1981):

$$D = 2(h + \sqrt{\left(\frac{\rho_s}{2\pi F\mu_0}\right)} \qquad (7)$$

where h is the height of the SWER line above ground, $\rho_s$ is the soil resistivity. To use this for a SWER line we replace the line separation 2h in equation (6) by D in equation (7).

At 50 Hz the return path is typically 1km below ground and at 100 kHz it is about 36 m below ground. In both cases the return path distance is still much less than a quarter wavelength, so no additional nulls appear in the field from the elementary doublet. Since the effective line spacing varies, the line capacitance and characteristic impedance also change with frequency.

Equation (6) can also be used to determine the radiation from overhead power lines where the PLC signals are applied in a differential form on two conductors with a line separation of 2h.

From knowing the skin depth, the diameter of the conductor and its conductivity, the resistance of conductor for the elementary doublet length can be calculated. Knowing the current flowing through the elementary doublet, the resistive losses can then be calculated.

Knowing the current flowing into the present elementary doublet and keeping the characteristic impedance of the line fixed at the value for the operating frequency, the current $I_0$ flowing into the next elementary doublet can determined by subtracting the cumulated power lost in radiation and the cumulated resistive losses from the input power to the present elementary doublet as shown in (8).

$$I_0 = \sqrt{\frac{I_{in}^2 Z_0 - P_{rad} - P_{res}}{Z_0}}$$
(8)

This process of adding an elementary doublet, calculating the resulting cumulative $E_\theta$, cumulative radiated power, resistive losses and evaluating the current $I_0$ flowing into the next elementary doublet, is repeated to determine the $E_\theta$ and the power radiated and the resistive losses for the whole line. The accuracy of this model has been verified by the authors (Reid & Kikkert, 2008) by comparing its results for short line lengths with those from standard equations.

### 2.2 Transmission line parameters

To calculate the resistive losses and radiation losses, accurate transmission line parameters need to be used. The fundamental equations for transmission line parameters are shown in (Johnson, 1950). Those equations have been used by the author (Kikkert & Reid, 2009) to derive the line parameters for power lines.

In a conductor, most of the current flows in the outer layer of the conductor, with a thickness of the skin depth ($\delta$). The equation for the resistance of the line is given by equation (9) (Johnson 1950):

$$R = \frac{\rho}{\sqrt{2}\pi a \delta} \frac{ber(q)\dfrac{d}{dq}bei(q) - bei(q)\dfrac{d}{dq}ber(q)}{\left\{\dfrac{d}{dq}bei(q)\right\}^2 + \left\{\dfrac{d}{dq}ber(q)\right\}^2} \quad \Omega / m$$
(9)

where *ber* and *bei* are Kelvin-Bessel functions (Johnson 1950), $a$ is the conductor radius, $\rho$ is the resistivity of the conductor in ohm-meters, $q$ is an intermediate constant and $\delta$ is the skin depth, $F$ is the frequency in Hz and $\mu$ is the magnetic permeability (4π x 10$^{-7}$ in free space), as given by equations (10) and (11).

$$q = \sqrt{2}\frac{a}{\delta} \qquad \delta = \sqrt{\frac{\rho}{\pi \mu F}}$$
(10)

$$R_{dc} = \frac{\rho}{\pi a^2} \quad \Omega / m$$
(11)

At high frequencies Fig 3.10 of (Johnson, 1950) shows equation (9) becomes approximately:

$$R = R_{dc}\frac{a}{2\delta}$$
(12)

To avoid calculation Kelvin-Bessel functions and considering that the conductor is made up of several strands with a steel core and aluminium cladding, equation (9) can be approximated to:

$$R = \sqrt{\frac{R_{Cdc}{}^2 a_s{}^2}{4\delta} + R_{Wdc}{}^2} \qquad (13)$$

where $R_{cdc}$ is the DC resistance of the whole conductor, if it were made from aluminium only, $R_{Wdc}$ is the actual DC resistance of the whole conductor and $a_s$ is the radius of each strand. At high frequencies the resistance is determined by the skin depth in the aluminium cladding and the low frequency asymptote is the actual DC resistance of the whole steel core bundled conductor. If a conductor is made up of solid steel or solid aluminium, then $R_{Cdc} = R_{Wdc}$. The difference between (9) and (13) for Steel Cored Galvanized Zinc (SCGZ) conductors is less than 3.4% for frequencies above 3 kHz. For Aluminium conductors, the error is less than 3.4% for frequencies above 30 kHz. Equation (9) cannot be applied to SCAC conductors, so that equation (13) is an acceptable equation for all conductors at PLC frequencies.

The line inductance for a line, can similarly be approximated to:

$$L_i = \frac{\mu_r \mu_0}{\sqrt{\dfrac{a^2}{4\delta^2} + 1}} \qquad (14)$$

$$L = \frac{4}{10K} Log_e\left(\frac{D}{a} + L_i\right) \ mH / km \qquad (15)$$

where $L$ is the inductance of the line, $L_i$ is the internal inductance, $D$ is the distance between the conductor and the one carrying the return path. For SWER lines, this is given by equation (7). $K$ is a constant, where $K = 2$ if the transmission line consists of a conductor above a ground plane and $K = 1$ if the transmission line consists of a balanced line.

$$C = \frac{2K\varepsilon_r}{36 Log_e(\dfrac{D}{a})} \ \mu F / km \qquad (16)$$

$$Z = \sqrt{\frac{L}{C}} \qquad (17)$$

## 3. Matlab® implementation

### 3.1 Transmission line parameters

For SWER lines, three types of conductors are commonly used. The oldest lines use Steel Core Galvanised Zinc (SCGZ) cables. These are basically fencing wire and typically have 3 strands, each of 2.75 mm diameter (Olex, 2006). The DC resistance at 20 °C is 11 Ω. The zinc coating is very thin and it's thickness is not adequately specified and varies with age. As a result these lines are modelled as steel lines.

Most SWER lines have a high current backbone, which then feeds many lower current spurs. The newer spurs use Steel Core Aluminium Clad (SCAC) cables. Typically these cables have 3 strands, each of 2.75 mm diameter (Olex, 2006), with an aluminium area of 5.91 mm² and a total area of 17.28 mm². The DC resistance at 20 °C is 4.8 Ω. The high power backbone typically uses Sultana (Olex, 2006) cables, which consist of 4 strands of Aluminium wires of 3 mm diameter around a galvanised steel core of 3 wires of 3 mm diameter. The DC resistance at 20 °C is 0.897 Ω. Since the conductivity of the steel is much less than the aluminium, these lines are modelled as pure aluminium conductors. Other power line conductor configurations are used in practice. They can be incorporated by simply specifying the correct wire diameter, the diameter of the whole conductor, the number of strands and for composite conductors, the area of Aluminium and the area of steel. In addition the conductivity and the permeability of the cable is also required. Those variables are listed at the start of the Matlab® program, the relevant ones for a SCAC conductor are as follows:

RhoC= 2.82e-8;       %resistivity of conductor 2.82e-8 for Al, 1.74e-7 for steel
RhoS= 30.0 ;       %resistivity of soil typical 30
DiaC= 2.75e-3;       %diameter of conductor strand in m for default 2.75mm
ALArea=5.91;       %Area of Aluminium in conductor SCAC
LineArea=17.82;       %Conductor Area SCAC
NSC=3;       %number of conductor strands default 3
DiaL=5.9e-3;       %Nominal Line Diameter m, 5.9 mm for SCGZ, SCAC
Er=4.5;       %Dielectric constant for the spacing between conductors. 4.5 for SWER from measurements
RadC = DiaC/2.0;       %conductor radius for skin depth and resistance
RadL = DiaL/2.0;       %Line radius for capacitance and inductance
Mu0=4.0*pi*1.0e-7;       %mu0 for free space
MuR=1.0;       % Mur=70 for steel (SCGZ), 1 for Aluminium (SCAC and Sultana)
MuC = Mu0*MuR;
Rdc = RhoC*1000.0/(NSC*pi*RadC*RadC);
RdcW = Rdc*LineArea/ALArea ;       %DC resistance of Aluminium part of the line.

For SCGZ conductors, the above values are changed to the SCGZ values. For SCGZ and pure aluminium conductors, the last line above is simply changed to:

RdcW = Rdc;       %DC resistance of Aluminium part of the line or Steel for a SCGZ line.

The Matlab® program uses these line parameters to calculates the skin depth using equation (10) and uses that together with equations (11) and (13) to determine the actual line resistance. The line inductance and characteristic impedance of the line is calculated using equations from (Johnson, 1950) and Carson's Depth equation (7) as needed. The transmission line impedance, is then used to determine the resistive losses of the transmission line and using equation (8) is used to determine the current flowing into the next elementary doublet. The relevant equations are:

SkinDepth = sqrt(2.0*RhoC/(twopi*Freq*MuC));       %Johnson 1950, pp 74
Rs = (sqrt(Rdc*Rdc*RadC*RadC/(4.0*SkinDepth*SkinDepth)+RdcW*RdcW));
    %Line resistance per km asymptote at actual DC resistance of line (RdcW).
Lr = 1.0/(sqrt(RadL*RadL/(4.0*SkinDepth*SkinDepth)+1));       % internal inductance
if FreeSpace == 'y';       %if 'y' Line in Free Space if 'n' line with groundplane or balanced transmission line.

```
    RealZo = Mu0*c;    %Single line characteristic impedance in free space 376 ohm
  else
  if SWER == 'y';  %if 'y' SWER Line, use Carson for line separation, if 'n' have balanced line
      K1=2.0;  %for SWER K1=2 Line with Groundplane, like SWER line
      LineSep= 2.0*(h+sqrt(RhoS/(twopi*Freq*Mu0)));  %Carson's depth
      Cl = Er*K1/(36.0*log(LineSep/RadL));  %Line Capacitance nF/m SWER line
    else
      K1=1.0;  %for SWER K1=2 Balanced line.
      LineSep = 2.0*h;  %Line separation for differential line configuration
      Cl = Er/(36.0*log(LineSep/RadC));  %Line capacitance nF/m = uF/km Line in free
  space
    end;
    Lkm = (4*log(LineSep/RadL)+MuR*Lr)/(K1*10);  %inductance (mH)/km
    Ckm = Cl*1000.0;  %line capacitance per km
    RealZo = sqrt(1.0e6*Lkm/Ckm); % Real part of Z for resistive losses.
  end;
```

## 3.2 Radiated power

The evaluation of equation (6) is the core of the Matlab® program that is used to calculate the resulting $E_\theta$ fields for these cascaded line segments, using (2) to (3) and the intermediate equations shown in Reed & Kikkert, 2008. For a single line in space,  the field is constant with respect to $\varphi$ and equations (2) to (5) can be used to determine the radiation pattern and the radiated power. For a line above ground, equation (6) needs to be used and integration over both $\varphi$ and $\theta$ need to be done. A 10 km line at 100 kHz is 3.33 wavelengths long and as shown in figure 1, has 14 main lobes.

Doubling the signal frequency, doubles the electrical length of the line and thus the number of lobes in the radiation pattern, and requires a halving of the angular interval for the determination of the radiated power for integrating equation (6) for E over $\theta$ and $\varphi$. In addition, the line has twice the number of elementary doublets, thus requiring 8 times number of calculations to be done. It is thus critical to get the optimum electrical line length for the elementary doublet and the optimum angular increment to obtain good accuracy, while minimising the number of calculations required. The initial work for this topic was done using a elementary doublet length of 0.01 λ or 0.1 km, whichever is the smallest, to ensure a high level of accuracy, so that for a 10 km line at 100 kHz, 299 line segments are used and at 100 MHz, 299980 line segments are used. For the results presented here the length for the elementary doublet is chosen to be 1/64 wavelength or 0.1 km, whichever is the smallest, since simulations showed that reducing the length below the 1/64 wavelength starts to affect the accuracy and 0.1 km is required since the printing of results are done at 0.1 km intervals. This then results in 200 elementary doublets for the line of figure 1.

The initial work for this topic was done at the 3600 angular intervals regardless of the signal frequency, to ensure that accurate fields were obtained for a 100 km long line at 100 kHz. It was found that one quarter of the number of elementary doublets used results in an accurate measure of radiated power, resulting in 50 angular intervals. The integration of the fields in the θ direction is done over the whole 0 to 360 degree region, and those in the φ direction only need to be done if from 0 to 90 degrees, due to symmetry. If the angular increment in the θ and φ directions are to be the same, then the number of angles is to be divisible by 4.

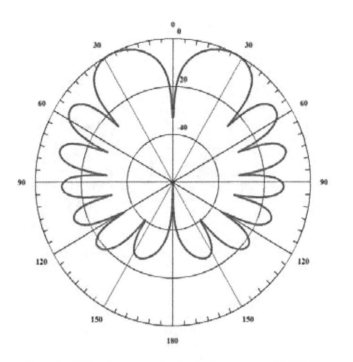

Fig. 1. Radiation pattern for 10 km long conductor in free space at 100 kHz.

As a result 52 angular intervals are required. This gives the same accuracy to 5 decimal places compared with 3600 angular intervals. To produce nice looking plots, as shown in figure 1, a minimum of 256 angles are used. The relevant code is shown below:

```
% N is the number of elementary douplets making up the
NumTh=4*round(N/16);   %number of angles per circle same as N by 4 as default.
if (NumTh > 3600) NumTh = 3600; end; %Limit to 0.1 degree.
if (NumTh < 256 ) NumTh = 256; end; %Limit to 256 for good looking plots
dtheta=twopi/NumTh;   %delta theta for a vector of angles and for integration 360 deg
```
For long power lines, the radiation pattern has many narrow lobes, as a result many angular intervals are required for the Matlab® evaluation of the radiation patterns. To avoid having to calculate sine and cosine for the same angles over and over again, these sine and cosine values are stored in an array, thus speeding up the calculations. This is done as follows:

```
% COMPUTE ANGLES, SINE OF ANGLES AND COSINE OF ANGLES
for angle=1:NumTh
    angles(angle)=dtheta*(angle-1) + 0.5*dtheta; %array of angles
    sinarr(angle)=sin(angles(angle));   %array of Sine used for both Theta and Phi
    cosarr(angle)=cos(angles(angle));   %array of Cos used for both Theta and Phi
end;
```
For a line in free space, this array is then used to calculate the cumulative E(θ) due to the elementary doublet length (dipL) having been added to E(θ) as shown below. This Esum(Theta) is the cumulative E(θ) and can then be used to calculate the cumulative radiated power as follows:

```
for segment=1:1:N   %N is number of segments
    if I>0
        for Theta=1:1:NumTh    %Loop to consider all angles Theta 0-360
            Eed_M(Theta) = 60*pi*edL*I*sinarr(Theta)*sin(EedPhi);
        %Field of elementary doublet of length edL, eqn (6), complex variable.
        end;
        edL_Ph=twopi_edL*(segment-1);   %Phase of I in current edL segment.
        for Theta=1:1:NumTh    %Loop to consider all angles Theta
            E_RE(Theta)= cos(edL_Ph*(cosarr(Theta)-1))*Eed_M(Theta); % real E_ field
            E_IM(Theta)=-sin(edL_Ph*(cosarr(Theta)-1))*Eed_M(Theta); % imag E_ field
            Etp(Theta)=complex(E_RE(Theta),E_IM(Theta)); %Total E field
            Esum(Theta)=Esumd(Theta)+Etp(Theta); %Progressive sum
            Esumd(Theta)= Esum(Theta);   %store current value for next iteration
        end;
        % Calculate the radiated & dissipated power and attenuation
        PowerRadiated=0;
        for Theta=1:1:NumTh
            E_squared(Theta)= (real(Esum(Theta))*real(Esum(Theta)) +
                imag(Esum(Theta))*imag(Esum(Theta)))';
            PowerRadiated = (PowerRadiated +E_squared(Theta)*abs(sinarr(Theta))
                *dtheta)/240);   %Integration 360 deg, 240 = 4x60
        end;
        Pdiss = Pdiss + I*I*ResistperSegment;
        PdissN = Pdiss/Pin; %Resistive Loss, normalise wrt input.
        PRadN = PowerRadiated/Pin; %Radiated power, normalise wrt input.
        %calculate current at output of dipole.
        I2 = (Pin - PowerRadiated - Pdiss)/RealZo; %Current into next segment
        if I2>0
            I = sqrt(I2);
        else
            I = 0;  %No more current left all dissipated.
        end;
        AttenuationResulting = 20*log10(I/Io); %cumulative total attenuation of the line
        Pout = I*I*RealZo/Pin;
    end; %Exit if no current left.
    % more code writing to file, plotting etc.
end;
```

For a line above a ground plane, a 3 dimensional integration, with $\theta$ and $\varphi$ must be made. Here the axis origin is at the groundplane and the maximum radiated field of the elementary doublet is perpendicular to the groundplane. The 3 dimensional field is calculated by adding another loop with a variable Phi, around the code for evaluation E($\theta$) as shown as follows:

```
if I>0
    for Phi=1:1:NumPhi   %Loop to consider all angles Phi (0-90)
        for Theta=1:1:NumTh    %Loop to consider all angles Theta 0-360
            Eed_M(Theta,Phi) = 60*pi*edL*I*sinarr(Theta)*sin(EedPhi(Phi)); % eqn (6)
```

```
      end;
    end;
    edL_Ph=twopi_edL*(segment-1);    %Phase of I in current edL segment.
    for Phi=1:1:NumPhi    %Loop to consider all angles Phi (0-90)
    for Theta=1:1:NumTh    %Loop to consider all angles Theta
        E_RE(Theta,Phi)= cos(edL_Ph*(cosarr(Theta)-1))*Eed_M(Theta,Phi); % real E
        E_IM(Theta,Phi)=-sin(edL_Ph*(cosarr(Theta)-1))*Eed_M(Theta,Phi); % imag E
       Etp(Theta)=complex(E_RE(Theta),E_IM(Theta));  %Total E field
       Esum(Theta,Phi)=Esumd(Theta,Phi)+Etp(Theta,Phi); %Progressive sum
       Esumd(Theta,Phi)= Esum(Theta,Phi);   %store current value for next iteration
      end;
    end;
    % Calculate the radiated & dissipated power and attenuation
    PowerRadiated=0;
    for Phi=1:1:NumPhi    %Loop to consider all angles Phi (0-90)
      for Theta=1:1:NumTh
       E_squared(Theta,Phi)= (real(Esum(Theta,Phi))*real(Esum(Theta,Phi)) +
         imag(Esum(Theta,Phi))*imag(Esum(Theta,Phi)))';
       PowerRadiated = (PowerRadiated + E_squared(Theta,Phi)*abs(sinarr(Theta))
         *dtheta)/240);        %Integration 360 deg 240 = 6x60
      end;
     if Phi == 1
         PowerRad2D=PowerRadiated; % same as for 2D plot.
       end;
    end;
     PowerRadiated=2*PowerRadiated*dtheta/pi; %note dtheta=dphi 3D radiated power
    RadN2D = PowerRad2D*(0.5+0.65*power(LineSepe,1.8))/Pin;
       % derived from 2D plot, same value
       % as 3D plot. allows 2D calculations to be used for lines with groundplane.
     Pdiss = Pdiss + I*I*ResistperSegment;
     PdissN = Pdiss/Pin; %Resistive Loss, normalise wrt input.
     PRadN = PowerRadiated/Pin; %Radiated power, normalise wrt input.
     %calculate current at output of dipole.
     I2 = (Pin - PowerRadiated - Pdiss)/RealZo; %Current into next segment
     if I2>0
         I = sqrt(I2);
     else
         I = 0;   %No more current left all dissipated.
     end;
     AttenuationResulting = 20*log10(I/Io); %cumulative total attenuation of the line
     Pout = I*I*RealZo/Pin;
   end; %Exit if no current left.
   % more code writing to file, plotting etc.
```

The 3D calculation for a 10 km SWER line with a 100 kHz PLC signal takes a few minutes on a fast PC. The same calculations for a 300 kHz PLC signal take several hours and the calculations for a 1 MHz PLC signal are expected to take several weeks, since less than 1% of the calculations were done overnight. It is thus nearly impossible to evaluate the fields from long lines at high frequency. Since the separation between the line and the return path is far less than a quarter wavelength for signals below 1 MHz, and as a result the dependence of the E fields has little dependence on φ. The radiated power derived from the field along the line and perpendicular to the ground plane can be used as an accurate measure of the radiated power of the power line. The accurate 3 D radiated power was calculated for many different frequencies, line lengths and line heights above ground. As a result, the radiated power for all these variations can accurately be determined as follows:

PowerRadiated = PowerRad2D*(0.5+0.65*power(LineSepe,1.8));

PowerRad2D is the power obtained from the 2D field, LineSepe is the line separation in wavelengths. When the line separation is less than half a wavelength, the error due to the above approximation is better than 0.01%. A 2D calculation for the radiated power can thus be used reducing the calculation time by several orders of magnitude without loss of accuracy. The 3D radiated power is thus:

$$P_{rad}(\theta,\varphi)_{rms} = P_{rad}(\theta)_{rms} \times (0.5 + 0.65 * \left(\frac{2h}{\lambda}\right)^{1.8}) \tag{18}$$

As a result of (18) the above Matlab® code can also be used for long lines above ground, or to find the radiation of overhead power lines with the PLC signals driven on two conductors in a balanced fashion.

The main Matlab® program is over 500 lines long, but only the core parts of the code have been described in this section.

## 4. Wobble

Figure 2 shows the field pattern from a 100 km long straight line, with the left figure being in free space and the right figure obtained from a SWER line 7 m above ground. The maximum gain for the free space line is 52.9 dB, while the maximum gain for the SWER line

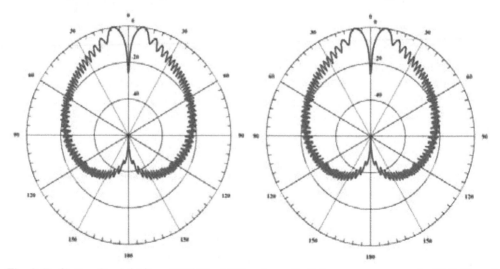

Fig. 2. Radiation Straight line, 100 kHz, 100 km long. Left free space, max gain 52.9 dB and Right SWER line max gain 21.6 dB.

is 21.6 dB. The radiation patterns look the same, so that the SWER line has 31.4 dB less peak radiated power than the free space line. The total radiated power from the SWER line is 32.4 dB less for the SWER line than the free space line.

Practical power-lines are not absolutely straight. Any small bends, as the line direction changes to follow the terrain, has a major impact on the radiation pattern. Figure 3 shows the effect of the radiation loss as the change in line direction, called *wobble* in this paper, is varied every 0.1 km by a random angle with the standard deviation indicated. Figure 3 is for a 1 km long SWER line with a 1 MHz carrier frequency. If the wobble is comparable to the spacing between the side lobes, large changes in attenuation can result.

For the radiation results presented in the rest of this chapter, a 50 mrad wobble is used as that reasonably represents the power line geometry, has sufficient randomness and results in a sufficient increase in radiation, while avoiding the rapid changes in attenuation associated with a larger wobble. Including the wobble, results in a significantly increase in radiated power. The average results for figure 3 are obtained from 10 simulations at each value of wobble.

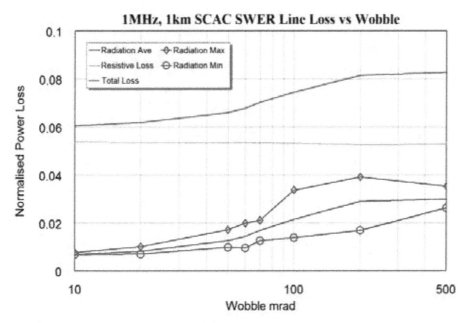

Fig. 3. Radiation loss of a SWER line with bends.

Figure 4 shows the radiated power for a line in free space and a SWER line when wobble is applied. The right graph of figure 2, shows the radiation patterns of a 100 km long straight SCAC SWER line with a 100 kHz carrier frequency, the right graph of figure 4 shows the corresponding radiation pattern of a SWER line with a 50 mrad wobble. The line with wobble has a 0.0415% radiation loss and the straight line has a 0.0118% radiation loss. The wobble will thus increase the radiated power by close to 4 times. The free space straight line has a radiation loss of 20.7%, while the line with wobble has lost all its input power after 90.5 km with a radiation loss of 60% and a resistive loss of 40%. Figure 5a shows these radiation losses versus distance for these four lines. It is interesting to note that for very long straight lines, the radiated power decreases as the line becomes longer. This is due to some of the fields cancelling as new elementary doublets are added. Figure 5 shows that the wobble increases the radiated power significantly and on average, the radiated power keeps increasing as the distance increases. For short lines, less than 3 km, the lobes of radiation are broad and the wobble is small compared to the beamwidth, so that the straight line and the line with wobble result in virtually the same radiation. For long lines the wobble is comparable to or larger than the beamwidth of the radiation pattern lobes and the difference between the radiation from the straight line and the line with wobble becomes larger.

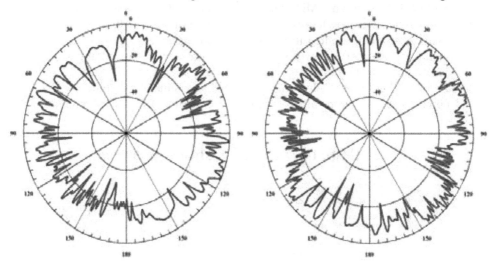

Fig. 4. Radiation line with 50 mrad wobble, 100 kHz, 100 km long. Left free space, max gain 49.8 dB and right SWER line max gain 18.4 dB.

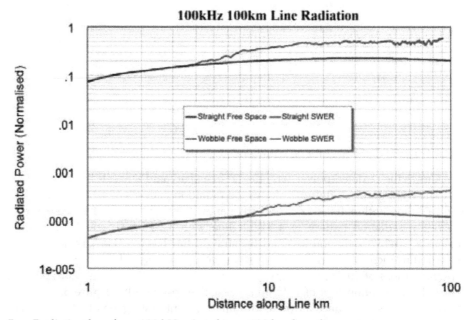

Fig. 5. a. Radiation loss for a 100 kHz signal on a 100 km long line.

Figure 5b shows the resistive loss corresponding to the radiation loss of figure 5a. Having a high radiation loss, results in a smaller current flowing along the line and thus less resistive ($I^2R$) losses. The radiation losses from SWER line are so small that the resistive loss with and without wobble is nearly the same.

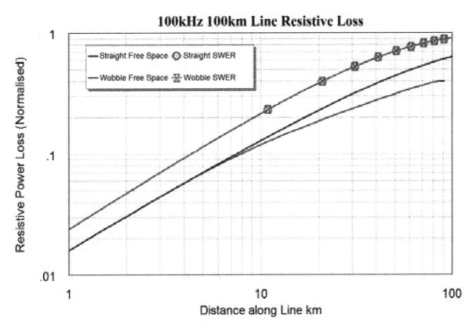

Fig. 5. b. Resistive loss for a 100 kHz signal on a 100km long line.

## 5. Results

### 5.1 SWER Lines

The first line type for which results are presented is a 3/2.75 SWER line consisting of either an aluminium clad steel (SCAC) or a galvanised steel (SCGZ) cable with three 2.75 mm diameter strands. The SWER line is typically 7 m above ground with the earth return distance varying according to equation (7). The author has made measurements on an 18 km long SCGZ SWER line SWER line (Kikkert & Reid 2009) at Mt Stewart station, 100 km west of Charters Towers in North Queensland, Australia. A comparison between the SWER line measurements and the results from the model presented here, using equations (7) to (17) and including the ground pad and earth return path resistance is shown in figure 6. It can be seen that there is a remarkable agreement, thus verifying the schematic model of the line using equations (7) to (17) and the corresponding Matlab® code in section 3.1.

The parameters for a 3/2.75 SCGZ SWER line obtained from the Matlab® program are shown in figure 7. The line resistance increases significantly with frequency. The earth return path depth (7) varies from close to 2 km at 10 Hz to 16 m at 10 MHz. Due to the current being carried in the steel conductor for SCGZ lines, the inductance varies with frequency. For aluminium or copper conductors, the internal inductance of equation (14) is negligible, so that the inductance is mainly governed by the external inductance. For steel conductors, with $\mu_r$ typically being 70, the internal inductance is larger, so that the inductance changes significantly with frequency. As a result of both of these effects, the characteristic impedance of the SWER line varies from 565 Ω at 50 Hz to 270 Ω at 100 kHz.

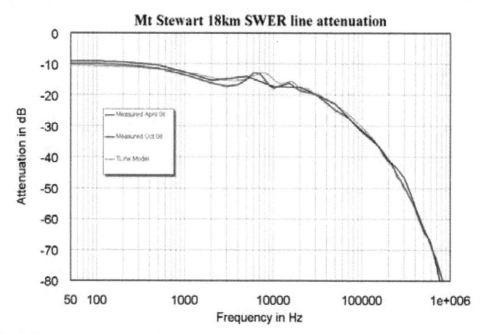

Fig. 6. Measurements and model comparison.

Figure 7 also shows the radiated power, expressed as a resistance. Below 300 kHz, the radiation is negligible, but at 5 MHz the same power is lost in radiation as in resistive losses. The radiation losses increase proportionally to the frequency[2.77]. Figure 8 shows the same plots for a SCAC conductor. Comparing figures 7 and 8 shows that the corner frequency where the line resistance starts to increase is much lower for the SCGZ conductors.

The parameters for a 3/2.75 SCAC SWER line obtained from the Matlab® program are shown in figure 8. The line resistance increases significantly with frequency. Despite the earth return path depth (7) varying from close to 2 km at 10 Hz to 16 m at 10 MHz, the characteristic impedance of the SWER line only varies from 344 Ω at 50 Hz to 258 Ω at 100 kHz. Figure 8 also shows the difference between the resistance of the line of it were a pure Aluminium line and the resistance of the SCAC line. Above 1 MHz all the current is carried in the Aluminium cladding. Figure 8 also shows the radiated power, expressed as a resistance. Below 300 kHz, the radiation is negligible, but at 1.5 MHz the same power is lost in radiation as in resistive losses. Equation (18) shows that the radiation losses increase proportionally to the frequency raised to the power 2.77. Comparing figures 7 and 8 show that the resistive losses of the SCGZ line are much higher than the SCAC line.

Figure 9 shows the line losses for a 3/2.75 SCGZ SWER line. For frequencies below 1 MHz, the radiation losses are very small in comparison to the resistive losses. For a to 10 km line and 100 kHz, the radiated power is 0.008% and for lower frequencies it is smaller. As a result radiation is negligible for all CENELEC-A (9 kHz to 95 kHz) frequencies. The resistive losses for SCGZ lines are such that for a 70 km SWER line a 33 dB attenuation results. This is still well below the typical dynamic range of PLC modems.

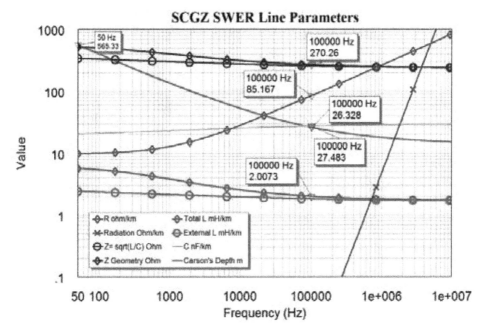

Fig. 7. Steel Core Galvanized Zinc (SCGZ) conductor per km transmission line parameters.

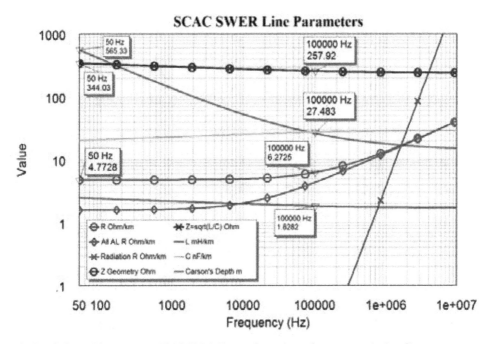

Fig. 8. Steel Core Aluminium Clad (SCAC) conductor per km transmission line parameters.

Figure 10 shows the corresponding line losses for a 3/2.75 SCAC SWER line. Comparing figures 9 and 10, show that for a 3/2.75 SCAC line the resistive losses are much lower. The radiation losses are the same, but the resistive losses are significantly smaller. A 300 km SCAC SWER line at 100 kHz has a 37 dB attenuation, similar to that of a 70 km SCGZ line.

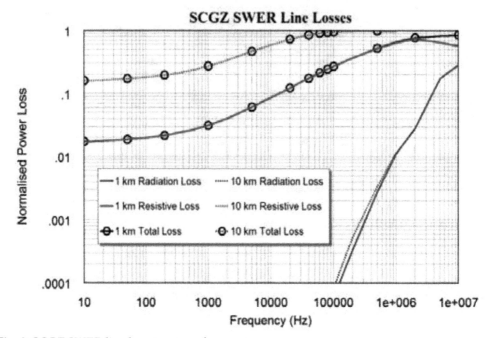

Fig. 9. SCGZ SWER line losses versus frequency.

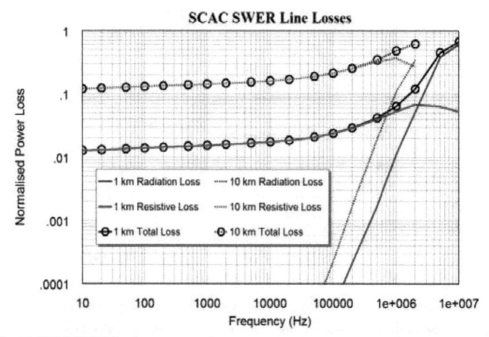

Fig. 10. SCAC SWER line losses versus frequency.

### 5.1.1 SWER line radiation resistance

The Matlab® program described in this chapter, prints the calculated normalised cumulative radiation and resistive losses every 100 m along the SWER line. Taking the difference between these cumulative results and averaging these for a line length of 30 km or when the total losses are less than half the input power, whichever is the shorter distance, allows a per km radiation resistance to be determined as a function of frequency. To ensure that the SWER line configuration is identical as the frequency is varied using different Matlab® program runs, the seed for the random wobble is initialised to the same value at the start of each run. Figure 11 shows the resulting radiation resistance for SCGZ and SCAC conductors for a SWER line 7 m above ground and a 50 mrad wobble every 100 m. Despite the resistive losses being very different, the per km radiation losses are nearly the same. Figure 11 also shows a least square error fit as follows:

$$R_{rad} = 1.35 \times 10^{-16} \times Frequency^{2.77} \quad \Omega / km \qquad (19)$$

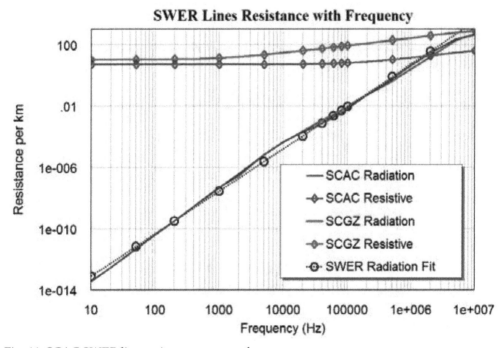

Fig. 11. SCAC SWER line resistances versus frequency.

Figure 11 shows that despite the coefficients of equation 21 being non-integer values, equation (19) gives a good approximation of the results obtained by the Matlab® program. At high frequencies, the radiation losses become quite high and within a short distance, all the input power is lost in resistive and radiation losses.

Figures 10 and 11 shows that for SCAC SWER line at 2 MHz, the radiation losses are equal to the resistive losses. For larger frequencies the radiation losses increase rapidly, causing a very high line attenuation. Since PLC transmissions by Electricity Utilities use frequencies below 150 kHz, figures 7 to 11, show that such PLC transmissions will not cause any interference as the radiated power is very small. However figures 7 to 11 show that SCAC SWER lines, SCGZ SWER lines and MV or LV three phase overhead power lines, where one phase is used for PLC and the earth path is used for the return path, are unsuitable for communication signals above 500 kHz, as the radiation losses will be significant, so that interference to other services in the MF band such as AM broadcast transmissions, may occur.

## 5.2 Overhead lines mounted on a crossarm

The second type of transmission line considered in this paper is the typical traditional open wire low voltage overhead (LVOH) line consisting of four All-Aluminium Conductor (AAC) cables each with 7 strands of 4.75 mm diameter mounted on a crossarm. The line consists of two sets of wires with a 550 mm spacing and a larger spacing between these sets of wires to accommodate the power pole. The results presented here are for a set of wires with a 550 mm spacing, which is typical in Australia for the outer sets of 240 V LVOH conductors on a crossarm.

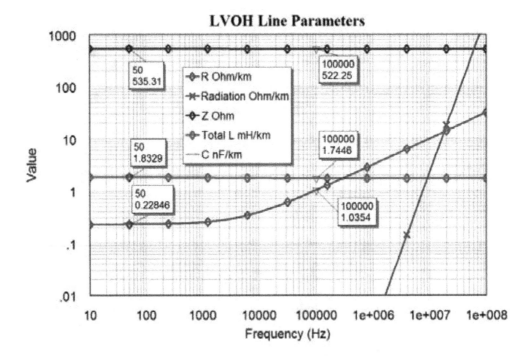

Fig. 12. LV overhead line on crossarm parameters versus frequency.

The parameters for the overhead line are shown in figure 12. The characteristic impedance of the line is 522 Ω and is virtually independent of frequency. There are only small variations of line inductance and capacitance with frequency. The plot for radiation in figure 12 corresponds to equation (20) and the OHL radiation curve fit in figure 14. For PLC applications, the signal is coupled onto these lines in a differential manner, and no ground return is used. Figure 13 shows that, the radiation losses of this line are much less than that of the SWER line.

Figure 14 shows the radiation and line resistance versus frequency for OHL and ABC lines as obtained from the Matlab® program described in this chapter. The least square error fit for frequencies above 1 kHz for OHL on crossarm conductors is:

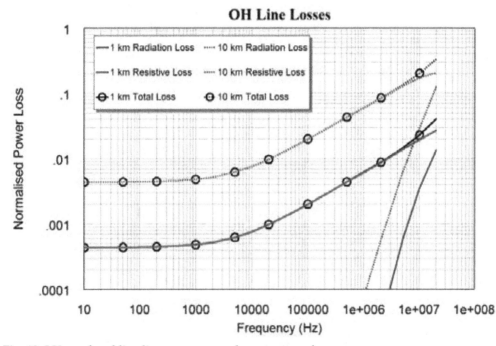

Fig. 13. LV overhead line line on crossarm losses versus frequency.

$$R_{rad} = 2.38^{-21} \times Frequency^3 \quad \Omega \, / \, km \tag{20}$$

The radiation resistance and the line resistance are the same at just above 20 MHz, so that these lines can be used at more than 10 times the frequency of SWER lines. If the PLC signal is coupled onto the all three phase lines in a common mode, i.e. the PLC signals is coupled onto all the lines in-phase, the neutral line is not connected to the PLC signal and the ground is used for a return path, then similar radiation losses to SWER lines are obtained. For a 10 km line, at 2 MHz, less than 0.05% of the input power is radiated and the line can thus be used for transmitting communication signals at those frequencies.

Broadband over Power Lines (BPL) uses signals in the frequency Range 525 kHz to 80 MHz. Field trials (ACMA 2007, ACMA 2009) have been carried out in Australia. The tests at Queanbean (ACMA2009) used frequencies in the range of 2.5 MHz to 22.8 MHz. Figure 13 shows that at 20 MHz and a 1 km line length, 15% of the power is lost in radiation while 25% is lost in resistive losses. BPL is thus not a suitable system to be used on LVOH lines. PLC signals using the CENELEC A band of 35 to 95 kHz for smart metering can be applied on these lines without any significant interference to existing radio services.

### 5.3 Aerial bundled conductors

The final type of transmission line considered is an Aerial Bundled Conductor (ABC), consisting of four insulated, compacted aluminium cables grouped together with common insulation. Each conductor is made up of 19 strands, 2.52 mm in diameter. Each compacted

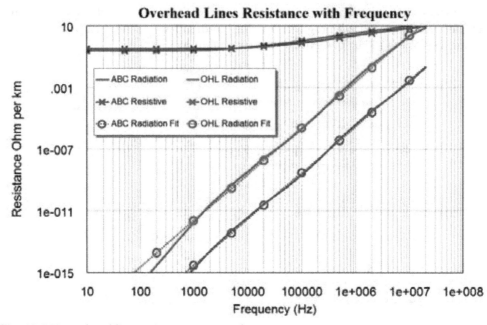

Fig. 14. LV overhead line resistances versus frequency.

conductor is 11.4 mm in diameter and the spacing between the centres of the conductors is 14.9 mm. The cable type is XDAB22AA004 from (Olex, 2008). The parameters of this line is shown in figure 15. The radiation resistance shown in figure 15 corresponds to equation (21) and the least curve fit for ABC in figure 14. For ABC lines, this corresponds to:

$$R_{rad} = 3.0^{-24} \times Frequency^3 \quad \Omega / km \qquad (21)$$

Comparing equations (19), (20) and (21), shows that for OHL on crossarm and ABC liones, the radiation resistance increases with the third power of frequency, while for SWER lines the radiation resistance increases with the frequency raised to the power 2.77. The difference is due to the earth return path and thus the effective line separation for SWER lines changing with frequency as shown by Carson's equation (7). Because the effective line separation is much larger for SWER lines, the radiation from SWER lines is much larger than those for the OHL and ABC lines.

Figure 16 shows the losses for ABC conductors for 1 km and 10 km ABC lines. The corresponding figures for the SWER and OHL on crossarm conductors are figures 9, 10 and 13. Comparing these figures show that ABC conductors have very little radiation.

Comparing the resistive, conductor, losses for OHL on crossarm and ABC conductors in figure 14, shows that at low frequency ABC conductors have a higher resistance than OHL on crossarm conductors but at high frequencies ABC conductors have a lower resistance. This is because the ABC conductors have 19 strands in the conductor, while the Moon conductor (Olex, 2008) used for the OHL on crossarm conductor has 7 strands. As a result the skin effect is less severe for ABC conductors. The calculations presented here have not considered that the 19 strands of the ABC conductors are compressed and that this will reduce the conductor surface available at high frequencies and thus increase the line resistance above the value calculated. Unfortunately the extent of this increase is not presented in any cable manufacturer's catalogue, since the main application for these lines is at 50 or 60 Hz.

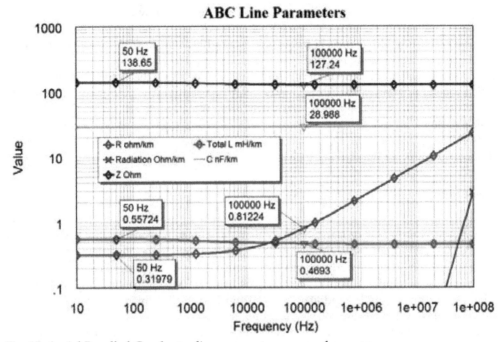

Fig. 15. Aerial Bundled Conductor line parameters versus frequency.

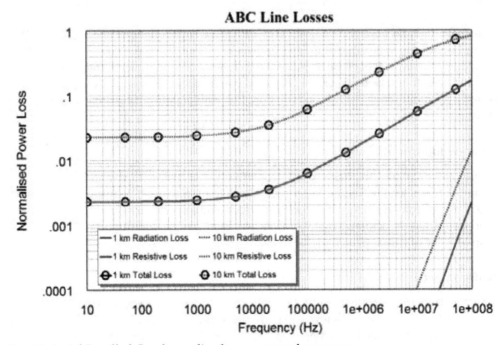

Fig. 16. Aerial Bundled Conductor line losses versus frequency.

Figure 16 shows that for frequencies below 100 MHz, the radiation losses are less than 2% of the input power. However the resistive losses are such that for frequencies above 20 MHz, all the input power is lost before the end of a 30 km line. The low level of radiation losses will ensure that no significant interference to other services will occur. ABC lines can thus be used for transmitting BPL signals at frequencies up to 100 MHz, however the resistive losses prevent great distances being reached.

Similar results will apply for underground cables, where all the active conductors are bundled together in one cable. BPL can thus safely be used on ABC lines and underground cables. The success of BPL on those lines will primarily depend on economic rather that technical matters.

PLC signals using the CENELEC A band for smart metering can be applied on these lines without any significant interference to existing services.

## 6. Conclusion

The radiation from power lines cannot be determined from expressions from long wire antennae, since those expressions assume that the signal current in the power line is constant, the power line is straight and its length is less than 5 wavelengths. None of these assumptions are true for powerlines and as a result the radiation patterns and the radiated power must be determined from first principles using a Matlab® program. This program can be used to determine the radiated power from both SWER lines and overhead power lines where the PLC signals are coupled to 2 lines in a differential mode.

The results presented here show that SWER line result in little radiation for frequencies below 100 kHz, typical overhead lines mounted on a crossarm have little radiation for frequencies up to 2 MHz and bundled conductors have little radiation at frequencies up to 100 MHz.

All lines can thus be used for the 9-95 kHz CENELEC A frequency band to permit PLC communication for smart-metering and network control. Conventional overhead lines mounted on crossarms will cause significant interference levels for signal frequencies above 2 MHz and should not be used for BPL. Bundled conductors, ABC and underground, can be used for BPL applications, with little radiation, but with significant resistive losses at high frequencies.

## 7. Acknowledgment

The author would like to thank ERGON ENERGY for their support in providing the funding for this research. The author would also like to thank the technical and academic staff at James Cook University, for their assistance and encouragement in this work. The author would like to acknowledge Mr. Geoffrey Reid for his contribution to writing the Matlab code for calculating the radiated fields.

## 8. References

ACMA "ACA Field Report Country Energy BPL Trial Queanbeyan NSW – February 2005" ACA URL: http://www.acma.gov.au/webwr /lib284/queanbeyan %20trial %20feb202005%20final%20public%20 report.pdf., 23 October 2009

Balanis, A., Antenna theory: analysis and design. 2nd edition ed. 1997: John Wiley & Sons, Inc.

Walter, C.H., Travelling Wave Antennas. 1970 ed. 1965: Dover Publications. 429.

Ulaby, F.T., Fundamentals of Applied Electromagnetics. Fifth Edition. 1994: Pearson Prentice Hall. 456.

Johnson, W.C., Transmission Lines and Networks. International Student Edition, 1950, McGraw-Hill.

Reid, G. D. and C. J. Kikkert, "Radiation from a single wire earth return power line", EEEvolution Symposium, Cairns 28-30 July 2008

Kikkert, C, J, and G. Reid, Radiation and Attenuation of Communication Signals on Power Lines. 7th International Conference on Information, Communications and Signal Processing, ICICS 2009, 7-10 December 2009, Macau.

Carson, J.R., Wave propagation in overhead wires with ground return. Bell Systems Technical Journal, 1926. 5.

Wang,Y., Liu, S. "A Review of Methods for Calculation of Frequency-dependent Impedance of Overhead Power Transmission Lines." 2001 Proc. Natl. Sci. Counc. ROC(A) Vol 25, No. 6, 2001. pp. 329-338

Deri, A.,Tevan, G., "The Complex Ground Return Plane: A Simplified Model for Homogeneous and Multi-layer Earth Return." IEEE Transactions on Power Apparatus and Systems, Vol. PAS-100, No. 8. August 1981.

Olex, Areal cable catalogue, March 2008 http://storage.baselocation.com/ olex.com.au/ Media/Docs/Aerial-Catalogue-b8c9b2b7-c2b0-42a2-a5a6-6c4d000f3df5.pdf

# 4

# PV Curves for Steady-State Security Assessment with MATLAB

Ricardo Vargas, M.A Arjona and Manuel Carrillo
*Instituto Tecnológico de la Laguna*
*División de Estudios de Posgrado e Investigación*
*México*

## 1. Introduction

Most of the problem solutions oriented to the analysis of power systems require the implementation of sophisticated algorithms which need a considerable amount of calculations that must be carried out with a digital computer. Advances in software and hardware engineering have led to the development of specialized computing tools in the area of electrical power systems which allows its efficient analysis. Most of the computational programs, if not all of them, are developed under proprietary code, in other words, the users does not have access to the source code, which limits its usage scope. These programs are considered as black boxes that users only need to feed the required input data to obtain the results without knowing anything about the details of the inner program structure. In the academic or research areas this kind of programs does not fulfill all needs that are required hence it is common the usage of programming tools oriented to the scientific computing. These tools facilitate the development of solution algorithms for any engineering problem, by taking into account the mathematical formulations which define the solution of the proposed problem. Besides, it is also common that most of these programs are known as script or interpreted languages, such as MATLAB, Python and Perl. They all have the common feature of being high level programming languages that usually make use of available efficient libraries in a straightforward way. MATLAB is considered as a programming language that has become a good option for many researchers in different science and engineering areas because of it can allow the creation, manipulation and operation of sparse or full matrices; it also allows to the user the programming of any mathematical algorithm by means of an ordered sequence of commands (code) written into an ascii file known as script files. These files are portable, i.e. they can be executed in most of software versions in any processor under the operating systems Linux or windows.

The main objective of this chapter consists in presenting an efficient alternative of developing a script program in the MATLAB environment; the program can generate characteristic curves power vs. voltage (PV curves) of each node in a power system. The curves are used to analyze and evaluate the stability voltage limits in steady state, and they are calculated by employing an algorithm known as continuous load flows, which are a variation of the Newton-Raphson formulation for load flows but it avoids any possibility of singularity during the solution process under a scenario of continuous load variation. To illustrate the application of this analysis tool, the 14-node IEEE test system is used to generate the PV curves. The code presented allows any modification throughout its script file and therefore it can be used for future power system studies and research.

The formulation of the load flow problem is firstly presented to obtain the PV curves, there are some issues that need to be taken into account in the algorithm oriented to the solution of load flows such as: mathematical formulation of the load flow, its adaptation to the Newton-Rhapson method and the implementation of the continuation theory to the analysis of load flows. It is also presented the necessary programming issues to the development of the script that plots the PV curves, the recommendations that are needed in the creation, manipulation and operation of sparse matrices, the use of vector operations, triangular decompositions techniques (that used in the solution of the set on linear equations) and finally the reading of ascii files and Graphical User Interface (GUI) development are also given.

## 1.1 Antecedents

Nowadays there are commercial programs which have been approved and used for the electric utilities in the analysis of electric power systems. Simulation programs as the Power World Simulator (PWS) (PowerWorld Corporation, 2010) and PSS (Siemens, 2005) are some of the most popular in the control and planning of a power system, and some of them are adopted by universities, e.g. PWS, because of its elegant interface and easy usage. Most of them have friendly user interfaces. On the other hand, a bachelor or graduate student, who want to reproduce or test new problem formulations to the solution of power system problems, need a simulation tool suitable for the generation of prototype programs. The code reutilization is important for integrating in a modular form, new functions required for the power system analysis (Milano, 2010). Commercial programs does not fulfill these requirements and therefore a search for alternatives is usually carry out, such as a new programming language or for the scientific language MATLAB. It is possible to find open source projects in several websites, which are usually named as "Toolbox" by their authors, and they cover a vast diversity of topics as: load flows, transient stability analysis, nodal analysis, and electromagnetic transients. Some of most relevant projects and its authors are: PSAT by Federico Milano (Milano, 2006), MatPower by Zimmerman, Carlos E. Murillo-Sánchez and Deqiang Gan (Zimmerman et al., 2011), PST by Graham Rogers, Joe H. Chow and Luigi Vanfretti (Graham et al., 2009) and MatEMTP by Mahseredjian, J. Alvarado and Fernando L. (Mahseredjian et al., 1997).

Similar projects have been developed at the Instituto Tecnológico de la Laguna (ITL) and they have been the basis for several MSc theses which have been integrated into the power system program PTL (figure 1). These projects have made possible the incorporation of new applications making a more flexible and robust program for the steady-state analysis of an electric power system.

## 2. Conceptual design of the PTL simulator

In spite of the foundation of the PTL program, i.e. being an integration of several graduate projects at the ITL, its design offers an interface which permits an intuitive user interaction and at the same time it has a dynamic performance which is able to solve load flow problems for any electric network regardless the node number. It offers the feature of showing the information graphically and numerically and besides it generates a report of the activities performed and exports files with data for making stability studies. The above PTL features make it a simulation program suitable for investigations because it allows

Fig. 1. GUI of the PTL program.

integrating solution algorithms for economic dispatch, calculation and plotting of PV curves, testing of methods with distributed slacks, static var compensator (SVC) models, transmission lines, generators, etc. It is also an important tool in power system research, making the PTL more complete for the analysis or studies oriented to the operation and control of electric power systems.

### 2.1 Data input details

As any other simulation program (commercial or free), the PTL requires of information data as input, it is needed for the analysis process. The information can be given as a data file that contains basic information to generate the base study case: the base power of the system, nodal information (number of nodes, voltages, load powers and generation power), machine limits, system branches, SVC information (if applies). The simulation program PTL can handle two file extensions: cdf (standardized IEEE format) and ptl (proposed PTL format).

### 2.2 Simulator description

The input information for carrying out the search of the solution process, as the related data to the problem results of the load flow problem (for generating the base case) are stored in defined data structures (e.g. Dat_Vn, Dat_Gen, Dat_Lin, Dat_Xtr). These structures allow easily the data extraction, by naming each field in such a way that the programming becomes intuitive and each variable can be easily identified with the corresponding physical variable of the problem. An example with two structures used in the PTL program is shown in Table 1.

The MATLAB structures are composed of non-primitive variable types that allow storing different data types in a hierarchical way with the same entity (García et al., 2005). They are formed by data containers called fields, which can be declared by defining the structure name and the desired field considering its value, e.g. Dat_Vn.Amp=1.02.

| Structure | Field | Description |
|-----------|-------|-------------|
| Dat_Vn | Amp | Voltage magnitude |
| | Ang | Phase angle of voltages |
| Dat_Gen | NumNG | Generation and load nodes |
| | Nslack | Slack node |
| | V_Rem_bus | Voltage information |
| | Pgen | Generated active power |
| | Qgen | Generated reactive power |
| | Pmin | Minimum limit of generator active power |
| | Pmax | Maximum limit of generator active power |
| | Qmin | Minimum limit of generator reactive power |
| | Qmax | Maximum limit of generator reactive power |

Table 1. Example of the information handled in the PTL.

### 2.3 Output information

The PTL program displays the results obtained from the load flow execution in a boxlist (uicontrol MATLAB) with a defined format: nodal information, power flows in branches and generators. It gives the option of printing a report in Word format with the same information. In addition, it has the option of generating a file with the extension f2s which is oriented for stability studies and it contains all necessary information for the initialization of the state machine variables by using the results of the current power flow solution.

The definition of the conceptual simulation program PTL is presented as a recommendation by taking into account the three basic points that a simulation program must include: to be completely functional, to be general for any study case and to facilitate its maintenance; in other words it must allow the incorporation of new functions for the solution of new studies, such that it allows its free modification as easy as possible.

## 3. Load flows

In a practical problem, the knowledge of the operating conditions of an electric power system is always needed; that is, the knowledge of the nodal voltage levels in steady-state under loaded and generating conditions and the availability of its transmission elements are required to evaluate the system reliability. Many studies focused to the electric power systems start from the load flow solution which is known as "base case", and in some cases, these studies are used to initialize the state variables of dynamic elements of a network (generators, motors, SVC, etc) to carry out dynamic and transient stability studies. Another study of interest, that it also requires starting from a base case, is the analysis of the power system security that will be discussed in next sections.

The mathematical equations used to solve this problem are known as power flow equations, or network equations. In its more basic form, these equations are derived considering the transmission network with lumped parameters under lineal and balanced conditions, similarly as the known operating conditions in all nodes of the system (Arrillaga, 2001).

### 3.1 Power flow equations

An electric power system is formed with elements that can be represented for its equivalent circuit RLC, and with components as load and generating units which cannot be represented as basic elements of an electrical network, they are represented as nonlinear elements. However, the analysis of an electrical power system starts with the formulation of a referenced nodal system and it describes the relationship between the electrical variables (voltages and currents) as it is stated by the second Kirchhoff´s law or nodal law.

$$\mathbf{I}_{BUS} = \mathbf{Y}_{BUS} \cdot \mathbf{V}_{BUS} \tag{1}$$

where $\mathbf{I}_{BUS}$ is a $n{\times}1$ vector whose components are the electrical net current injections in the $n$ network nodes, $\mathbf{V}_{BUS}$ is a $n{\times}1$ vector with the nodal voltages measured with respect to the referenced node and $\mathbf{Y}_{BUS}$ is the $n{\times}\,n$ nodal admittance matrix of the electrical network; it has the properties of being symmetric and squared, and it describes the network topology.

In a real power system, the injected currents to the network nodes are unknown; what it is commonly known is the net injected power $S_k$. Conceptually, $S_k$ is the net complex power injected to the $k$-th node of the electrical network, and it is determined by the product of voltage ($V_k$) and the current conjugate ($I_k^*$), where $V_k$ and $I_k$ are the voltages and nodal currents at the node $k$, that is, the $k$-th elements of vectors $\mathbf{V}_{BUS}$ y $\mathbf{I}_{BUS}$ in (1). Once the $I_k$ is calculated using (1), the net complex power $S_k$ can be expressed as:

$$S_k = V_k \cdot I_k^* = V_k \cdot \left( \sum_{m=1}^{n} Y_{k,m} V_m \right)^*, \text{ for } k= 1,2,.....n \tag{2}$$

where $Y_{k,m}$ is the element $(k,m)$ of $\mathbf{Y}_{BUS}$ matrix in (1). $S_k$ can also be represented for its real and imaginary components such as it is shown in the following expression:

$$S_k = P_k + jQ_k \text{ , for } k= 1,2,.....n \tag{3}$$

where $P_k$ and $Q_k$ are the net active and reactive power injected at node $k$ of the system, respectively. They are defined as:

$$P_k = P_k^{Gen} - P_k^{Load} \tag{4}$$

$$Q_k = Q_k^{Gen} - Q_k^{Load} \tag{5}$$

where the variables $P_k^{Gen}$ and $Q_k^{Gen}$ represent the active and reactive powers respectively. They are injected at node $k$ for a generator and the variables $P_k^{Load}$ and $Q_k^{Load}$ represent the active and reactive Powers, respectively of a load connected to the same node.

By representing the nodal voltages in polar form, we have:

$$V_k = V_k e^{j\theta_k} = V_k \left( \cos\theta_k + j\sin\theta_k \right) \tag{6}$$

and each element of the admittance matrix $\mathbf{Y}_{BUS}$ as,

$$Y_{km} = G_{km} + jB_{km} \tag{7}$$

Using the above expressions in (2), it results,

$$S_k = V_k e^{j\theta_k} \cdot \left( \sum_{m=1}^{n} (G_{km} + jB_{km}) V_m e^{j\theta_m} \right)^* = V_k \cdot \left( \sum_{m=1}^{n} V_m (G_{km} + jB_{km})(cos\theta_{km} + j\sin\theta_{km}) \right)^*$$
$$\text{for } k= 1,2,.....n \tag{8}$$

where $\theta_{km} = \theta_k - \theta_m$. By separating the real and imaginary parts as it is suggested in (3), it is obtained the following,

$$P_k = V_k^2 G_{kk} + V_k \sum_{\substack{m=1 \\ m \neq n}}^{n} V_m (G_{km} Cos(\theta_k - \theta_m) + B_{km} Sin(\theta_k - \theta_m)) \text{ for } k= 1,2,.....n \tag{9}$$

$$Q_k = -V_k^2 B_{km} + V_k \sum_{\substack{m=1 \\ m \neq n}}^{n} V_m (G_{km} Cos(\theta_k - \theta_m) - B_{km} Cos(\theta_k - \theta_m)), \text{ for } k= 1,2,.....n \tag{10}$$

The equations (9) and (10) are commonly known as Power flow equations and they are needed for solving the load flow problem (Arrillaga, 2001). By analyzing these equations it can be clearly seen that each system node $k$ is characterized for four variables: active power, reactive power, voltage magnitude and angle. Hence it is necessary to specify two of them and consider the remaining two as state variables to find with the solution of both equations.

## 3.2 Bus types in load flow studies

In an electrical power network, by considering its load flow equations, four variables are defined at each node, the active and reactive powers injected at node $P_k$ and $Q_k$, and the magnitude and phase voltage at the node $V_k$ y $\theta_k$. The latter two variables determine the total electrical state of the network, then, the objective of the load flow problem consists in determining these variables at each node. The variables can be classified in controlled variables, that is, its values can be specified and state variables to be calculated with the solution of the load flow problem. The controlled or specified variables are determined by taking into account the node nature, i.e. in a generator node, the active power can be controlled by the turbine speed governor, and the voltage magnitude of the generator node can be controlled by the automatic voltage regulator (AVR). In a load node, the active and reactive power can be specified because its values can be obtained from load demand studies. Therefore, the system nodes can be classified as follows:

- *Generator node PV*: It is any node where a generator is connected; the magnitude voltage and generated active power can be controlled or specified, while the voltage phase angle and the reactive power are the unknown state variables (Arrillaga, 2001).
- *Load node PQ*: It is any node where a system load is connected; the active and reactive consumed powers are known or specified, while the voltage magnitude and its phase angle are the unknown state variables to be calculated (Arrillaga, 2001).
- *Slack node* (Compensator): In a power system at least one of the nodes has to be selected and labeled with this node type. It is a generating node where it cannot be specified the generated active power as in the PV node, because the transmission losses are not known beforehand and thus it cannot be established the balance of active power of the loads and

generators. Therefore this node compensates the unbalance between the active power between loads and generating units as specified in the PQ and PV nodes (Arrillaga, 2001).

### 3.3 Solution of the nonlinear equations by the Newton-Rhapson method

The nature of the load flow problem formulation requires the simultaneous solution of a set of nonlinear equations; therefore it is necessary to apply a numerical method that guarantees a unique solution. There are methods as the Gauss-Seidel and Newton-Rhapson, in the work presented here, the load flow problem is solved with the Newton-Raphson (NR) (Arrillga, 2001).

The NR method is robust and has a fast convergence to the solution. The method has been applied to the solution of nonlinear equations that can be defined as,

$$\mathbf{f}(\mathbf{x}) = \mathbf{0} \tag{11}$$

where $\mathbf{f}(\mathbf{x})$ is a $n \times 1$ vector that contains the $n$ equations to be solved $f_i(\mathbf{x}) = 0$, $i=1,n$,

$$\mathbf{f}(\mathbf{x}) = \left[ f_1(\mathbf{x}), f_2(\mathbf{x}), \ldots, f_n(\mathbf{x}) \right]^T \tag{12}$$

and $\mathbf{x}$ is a $n \times 1$ vector that contains the state variables, $x_i$, $i=1,n$,

$$\mathbf{x} = \left[ x_1, x_2, \ldots, x_n \right]^T \tag{13}$$

The numerical methods used to solve (11) are focused to determine a recursive formula $\mathbf{x}^{k+1} = \mathbf{x}^k + \Delta \mathbf{x}$. The solution algorithm is based in the iterative application of the above formula starting from an initial estimate $\mathbf{x}^0$, until a convergence criterion is achieved $\max(\Delta \mathbf{x}) < \varepsilon$, where $\varepsilon$ is a small numerical value, and the vector $\mathbf{x}^k$ is an approximation to the solution $\mathbf{x}^*$ of (11).

The Newton-Raphson method can be easily explained when it is applied to an equation of a single state variable (Arrillaga, 2001). A geometrical illustration of this problem is shown in figure 2.

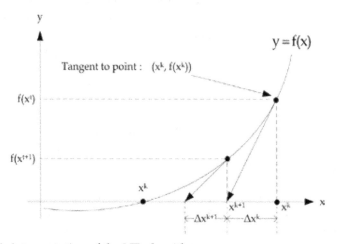

Fig. 2. Geometric interpretation of the NR algorithm.

From figure 2, it can be established that $\Delta x^k = x^{k+1} - x^k$

$$d\left.\frac{f(x)}{dx}\right|_{x^k} = -\frac{f(x)}{\Delta x^k} \tag{14}$$

where,

$$\Delta x^k = -\left(f'(x_k)\right)^{-1} \cdot f(x_k) \tag{15}$$

As it can be seen in figure 2, the successive application of this correction $\Delta x^k, \Delta x^{k+1}, \Delta x^{k+2}...$ leads to the solution $\mathbf{x}^*$ as nearer as desired.

To derive the recursive formula to be employed in the solution of the set of equations, and by expressing the equation (15) in matrix notation, it is obtained,

$$\Delta \mathbf{x}^k = -\left(\mathbf{f}'(\mathbf{x}_k)\right)^{-1} \cdot \mathbf{f}(\mathbf{x}_k) = -\left[\mathbf{J}(\mathbf{x}_k)\right]^{-1} \cdot \mathbf{f}(\mathbf{x}_k) \tag{16}$$

where

$$\mathbf{J}(\mathbf{x}_k) = \frac{\partial \mathbf{f}(\mathbf{x}_k)}{\partial \mathbf{x}} = \begin{bmatrix} \dfrac{\partial f_1(\mathbf{x})}{\partial x_1} & \dfrac{\partial f_1(\mathbf{x})}{\partial x_2} & \cdots & \dfrac{\partial f_1(\mathbf{x})}{\partial x_n} \\ \dfrac{\partial f_2(\mathbf{x})}{\partial x_1} & \dfrac{\partial f_2(\mathbf{x})}{\partial x_2} & \cdots & \dfrac{\partial f_2(\mathbf{x})}{\partial x_n} \\ \vdots & \vdots & \cdots & \vdots \\ \dfrac{\partial f_n(\mathbf{x})}{\partial x_1} & \dfrac{\partial f_n(\mathbf{x})}{\partial x_2} & \cdots & \dfrac{\partial f_n(\mathbf{x})}{\partial x_n} \end{bmatrix} \tag{17}$$

and the recursive formula is given by,

$$\mathbf{x}^{k+1} = \mathbf{x}^k + \Delta \mathbf{x} \tag{18}$$

### 3.4 Application of the NR to the power flow problem

The equations to be solved in the load flow problem, as it was explained in section 3.1, are here rewritten,

$$P_k^{esp} - \left[ V_k^2 G_{kk} + V_k \sum_{\substack{m=1 \\ m \neq n}}^{n} V_m \left( G_{km} Cos(\theta_k - \theta_m) + B_{km} Sin(\theta_k - \theta_m) \right) \right] = 0 \tag{4}$$

$$Q_k^{esp} - \left[ -V_k^2 B_{km} + V_k \sum_{\substack{m=1 \\ m \neq n}}^{n} V_m \left( G_{km} Cos(\theta_k - \theta_m) - B_{km} Cos(\theta_k - \theta_m) \right) \right] = 0 \tag{5}$$

for $k = 1, 2, \ldots n$

The $2n$ equations to be solved are represented by (4) and (5). However, for all generator nodes, equation (5) can be omitted and for the slack node the equations (4) and (5) (Arrillaga, 2001). The resulting set of equations is consistent because for the neglected equations, its corresponding state variables $P^{esp}$, $Q^{esp}$ are also omitted from them. All this operation gives as a result a set of equations to solve, where its state variables $x$ only contain magnitudes of nodal voltages and its corresponding phase angles which are denoted by $V$ y $\theta$, which simplifies considerably a guaranteed convergence of the numerical method. Therefore, the set of equation can be represented in vector form as,

$$\mathbf{f}(\mathbf{x}) = \begin{bmatrix} \Delta\mathbf{P}(\mathbf{x}) \\ \Delta\mathbf{Q}(\mathbf{x}) \end{bmatrix} = \mathbf{0} \tag{19}$$

where $\Delta\mathbf{P}(\mathbf{x})$ represents the equation (4) for PV and PQ nodes, $\Delta\mathbf{Q}(\mathbf{x})$ represents the equation (5) for PQ nodes and $\mathbf{x}$ denotes the state variables $\mathbf{V}$ and $\theta$, which are represented in vector notation as:

$$\mathbf{x} = \begin{bmatrix} \theta \\ \mathbf{V} \end{bmatrix} \tag{20}$$

where:
$\mathbf{V}$ = $nc$x1 vector.
$\theta$ = $nc+n$-1 vector.
$nc$ = Number of PQ nodes.
$n$ = Total number of nodes.
Considering equation (16), its matrix equation is obtained and it defines the solution of the load flow problem:

$$\underbrace{\begin{bmatrix} \Delta\theta \\ \Delta\mathbf{V} \end{bmatrix}}_{x^k} = - \underbrace{\begin{bmatrix} \dfrac{\partial\Delta\mathbf{P}}{\partial\theta} & \dfrac{\partial\Delta\mathbf{P}}{\partial\mathbf{V}} \\ \dfrac{\partial\Delta\mathbf{Q}}{\partial\theta} & \dfrac{\partial\Delta\mathbf{Q}}{\partial\mathbf{V}} \end{bmatrix}^{-1}}_{J(x^k)} \underbrace{\begin{bmatrix} \Delta\mathbf{P} \\ \Delta\mathbf{Q} \end{bmatrix}}_{f(x^k)} \tag{21}$$

By applying the recursive equation (18), the state variables ($V$ y $\theta$) are updated every iteration until the convergence criterion is achieved $\max(|\Delta\mathbf{P}(\mathbf{x})|) \le \varepsilon$ or $\max(|\Delta\mathbf{Q}(\mathbf{x})|) \le \varepsilon$ for a small $\varepsilon$ or until the iteration number exceeds the maximum number previously defined which in this indicates convergence problems.
The Jacobian elements in (21) are,

$$\frac{\partial\Delta\mathbf{P}}{\partial\theta} \tag{22}$$

They are calculated using (19), however, by taking into account that the specified powers $P_k^{esp}$ are constants, we obtain,

$$\frac{\partial\Delta\mathbf{P}}{\partial\theta} = -\frac{\partial\mathbf{P}}{\partial\theta} \tag{23}$$

Therefore, equation (21) can be expressed as,

$$\begin{bmatrix} \Delta \mathbf{P} \\ \Delta \mathbf{Q} \end{bmatrix} = \begin{bmatrix} \dfrac{\partial \mathbf{P}}{\partial \boldsymbol{\theta}} & \dfrac{\partial \mathbf{P}}{\partial \mathbf{V}} \\ \dfrac{\partial \mathbf{Q}}{\partial \boldsymbol{\theta}} & \dfrac{\partial \mathbf{Q}}{\partial \mathbf{V}} \end{bmatrix} \cdot \begin{matrix} \Delta \boldsymbol{\theta} \\ \Delta \mathbf{V} \end{matrix} \tag{24}$$

where the Jacobian elements are calculated using equations (9) and (10), provided equation (21) is normalized.

To simplify the calculation of the Jacobian elements of (24), it can be reformulated as:

$$\begin{bmatrix} \Delta \mathbf{P} \\ \Delta \mathbf{Q} \end{bmatrix} = \begin{bmatrix} \mathbf{H} & \mathbf{N} \\ \mathbf{J} & \mathbf{L} \end{bmatrix} \begin{bmatrix} \Delta \boldsymbol{\theta} \\ \dfrac{\Delta \mathbf{V}}{\mathbf{V}} \end{bmatrix} \tag{25}$$

where the quotient of each element with its corresponding element $V$ allows that some elements of the Jacobian matrix can be expressed similarly (Arrillaga, 2001). The Jacobian can be formed by defining four submatrixes denoted as **H**, **N**, **J** and **L** which are defined as,
If $k \neq m$ (off diagonal elements):

$$H_{km} = \frac{\partial P_k}{\partial \theta_m} = V_k V_m \left( G_{km} \sin \theta_{km} - B_{km} \cos \theta_{km} \right) \tag{26}$$

$$N_{km} = V_k \frac{\partial P_k}{\partial V_m} = V_k V_m \left( G_{km} \cos \theta_{km} + B_{km} \sin \theta_{km} \right) \tag{27}$$

$$J_{km} = \frac{\partial Q_k}{\partial \theta_m} = -V_k V_m \left( G_{km} \cos \theta_{km} + B_{km} \sin \theta_{km} \right) \tag{28}$$

$$L_{km} = V_k \frac{\partial P_k}{\partial V_m} = V_k V_m \left( G_{km} \sin \theta_{km} - B_{km} \cos \theta_{km} \right) \tag{29}$$

If $k = m$ (main diagonal elements):

$$H_{kk} = \frac{\partial P_k}{\partial \theta_k} = -Q_k - B_{kk} V_k^2 \tag{30}$$

$$N_{kk} = V_k \frac{\partial P_k}{\partial V_k} = P_k + G_{kk} V_k^2 \tag{31}$$

$$J_{kk} = \frac{\partial Q_k}{\partial \theta_m} = P_k - G_{kk} V_k^2 \tag{32}$$

$$L_{kk} = V_k \frac{\partial P_k}{\partial V_m} = Q_k - B_{kk} V_k^2 \tag{33}$$

It is important to consider that subscripts $k$ and $m$ are different, $L_{km} = H_{km}$ and $J_{km} = -N_{km}$, on the contrary, to the principal diagonal elements. Nevertheless, there is a relationship when the following equation is solved:

$$
\begin{bmatrix} V_1 & 0 & \cdots & 0 \\ 0 & V_1 & \cdots & 0 \\ \vdots & \vdots & \ddots & \vdots \\ 0 & 0 & \cdots & V_n \end{bmatrix} \left( \begin{bmatrix} Y_{11} & Y_{12} & \cdots & Y_{1n} \\ Y_{21} & Y_{22} & \cdots & Y_{2n} \\ \vdots & \vdots & \cdots & \vdots \\ Y_{n1} & Y_{n1} & \cdots & Y_{nn} \end{bmatrix} \begin{bmatrix} V_1 & 0 & \cdots & 0 \\ 0 & V_1 & \cdots & 0 \\ \vdots & \vdots & \ddots & \vdots \\ 0 & 0 & \cdots & V_n \end{bmatrix} \right)^* \tag{34}
$$

By simple inspection of (34), we can see that the operation for $-N_{km}$ and $J_{km}$ results to be the real part of (34), while $H_{km}$ and $L_{km}$ are the imaginary component of it. The variant consists that all elements of the principal diagonal are calculated after (34) has been solved by subtracting or adding the corresponding $P_k$ or $Q_k$ as it can be seen in (30)-(33).

The Jacobian matrix has the characteristic of being sparse and squared with an order of the length of vector $\mathbf{X}$, where its sub-matrixes have the following dimensions:

**H**: $n$-1 x $n$-1 matrix.

**N**: $n$-1 x $nc$ matrix.

**J**: $nc$ x $n$-1 matrix.

**L**: $nc$ x $nc$ matrix.

## 4. Voltage stability and collapse

The terms of voltage stability and collapse are closely related to each other in topics of operation and control of power systems.

### 4.1 Voltage stability

According to the IEEE, the voltage stability is defined as the capacity of a power system to maintain in all nodes acceptable voltage levels under normal conditions after a system disturbance for a given initial condition (Kundur, 1994). This definition gives us an idea of the robustness of a power system which is measured by its capability of keep the equilibrium between the demanded load and the generated power. The system can be in an unstable condition under a disturbance, increase of demanded load and changes in the topology of the network, causing an incontrollable voltage decrement (Kundur, 1994). The unstable condition can be originated for the operating limits of the power system components (Venkataramana, 2007), such as:

*Generators*: They represent the supply of reactive power enough to keep the power system in stable conditions by keeping the standardized voltage levels of normal operation. However, the generation of machines is limited by its capability curve that gives the constraints of the reactive power output due to the field winding current limitation.

*Transmission lines*: They are another important constraint to the voltage stability, and they also limit the maximum power that be transported and it is defined the thermal limits.

*Loads*: They represent the third elements that have influence on the stability voltage; they are classified in two categories: static and dynamic loads and they have an effect on the voltage profiles under excessive reactive power generation.

## 4.2 Voltage collapse

The voltage collapse is a phenomenon that might be present in a highly loaded electric power system. This can be present in the form of event sequence together with the voltage instability that may lead to a blackout or to voltage levels below the operating limits for a significant part of the power system (Kundur, 1994). Due to the nonlinear nature of the electrical network, as the phenomenon related to the power system, it is necessary to employ nonlinear techniques for the analysis of the voltage collapse (Venkataramana, 2007) and find out a solution to avoid it.

There many disturbances which contribute to the voltage collapse:

- Load increment.
- To reach the reactive power limits in generators, synchronous condensers or SVC.
- The operation of TAP changers in transformers.
- The tripping of transmission lines, transformers and generators.

Most of these changes have a significant effect in the production, consumption and transmission of reactive power. Because of this, it is suggested control actions by using compensator elements as capacitor banks, blocking of tap changers, new generation dispatch, secondary voltage regulation and load sectioning (Kundur, 1994).

## 4.3 Analysis methods for the voltage stability

Some of the tools used for the analysis of stability voltage are the methods based on dynamic analysis and those based in static analysis.

*Dynamic Analysis.* They consist in the numerical solution (simulation) of the set of differential and algebraic equations that model the power system (Kundur, 1994), this is similar as transients; however, this kind of simulations need considerable amount of computing resources and hence the solution time is large and they do not give information about the sensibility and stability degree.

*Static analysis.* They consist in the solution of the set of algebraic equations that represent the system in steady state (Kundur, 1994), with the aim of evaluating the feasibility of the equilibrium point represented by the operating conditions of the system and to find the critical voltage value. The advantage with respect to the dynamic analysis techniques is that it gives valuable information about the nature of the problem and helps to identify the key factors for the instability problem. The plotting of the PV curve helps to the analysis of the voltage stability limits of a power system under a scenario with load increments and with the presence of a disturbance such as the loss of generation or the loss of a transmission line.

## 4.4 PV curves

The PV curves represent the voltage variation with respect to the variation of load reactive power. This curve is produced by a series of load flow solutions for different load levels uniformly distributed, by keeping constant the power factor. The generated active power is proportionally incremented to the generator rating or to the participating factors which are defined by the user. The P and Q components of each load can or cannot be dependant of the bus voltage accordingly to the load model selected. The determination of the critical point for a given load increment is very important because it can lead to the voltage collapse of the system. These characteristics are illustrated in figure 3.

Some authors (Yamura et al, 1998 & Ogrady et al, 1999) have proposed voltage stability indexes which are based in some kind of analysis of load flows with the aim to evaluate the stability voltage limits. However, the Jacobian used in the load flows, when the Newton-Raphson is employed, becomes singular at the critical point, besides the load flow solutions at the points near to the critical region tend to diverge (Kundur, 1994). These disadvantages are avoided by using the method of continuation load flows (Venkataramana et al, 1992).

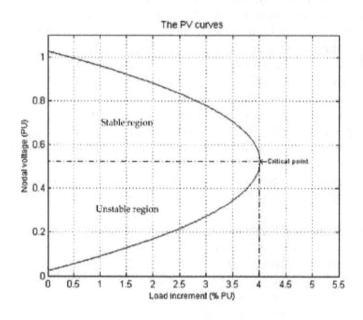

Fig. 3. PV curve

### 4.5 Application of the continuation method to the power flow problem
The continuous load flow procedure is based in a reformulation of the equations of the load flow problem and the application of the continuation technique with a local parameterization which has shown to be efficient in the trajectory plotting of PV curves.

The purpose of continuous load flows is to find a set of load flow solutions in a scenario where the load is continuously changing, starting from a base case until the critical point. Thereafter, the continuous load flows had been applied to understand and evaluate the problem of voltage stability and those areas that are likely to the voltage collapse. Besides, they have also been applied in other related problems like the evaluation of power transfer limits between regions.

The general principle of continuous load flows employs a predictor-corrector scheme to find a trajectory of solutions for the set of load flow equations (4) and (5) which are reformulated to include the load parameter $\lambda$.

$$\Delta P_i = \lambda \left( P_{Gi} - P_{Li} \right) - P_i = \lambda P_i^{esp} - P_i = 0 \tag{35}$$

$$\Delta Q_i = \lambda \left( Q_{Gi} - Q_{Li} \right) - Q_i = \lambda Q_i^{esp} - Q_i = 0 \tag{36}$$

$$1 \le \lambda \le \lambda_{crític} \tag{37}$$

The process is started from a known solution and a predictor vector which is tangent to the corrected solutions is used to estimate the future solutions with different values of the load parameter. The estimation is corrected using the same technique of the Newton-Rhapson employed in the conventional load flow with a new added parameter:

$$f(\theta, V, \lambda) = 0 \tag{38}$$

The parameterization plays an important role in the elimination of the Jacobian non-singularity.

### 4.5.1 Prediction of the new solution
Once the base solution has been found for $\lambda=0$, it is required to predict the next solution taking into consideration the appropriate step size and the direction of the tangent to the trajectory solution. The first task in this process consists in calculating the tangent vector, which is determined taking the first derivative of the reformulated flow equations (38).

$$d\mathbf{f}(\theta, V, \lambda) = \mathbf{f}_\theta d\theta + \mathbf{f}_V dV + \mathbf{f}_\lambda d\lambda = 0 \tag{39}$$

where F is the vector [ΔP, ΔQ, 0] that is augmented in one row; in a factorized form, the equation is expressed as,

$$\begin{bmatrix} \mathbf{f}_\theta & \mathbf{f}_F & \mathbf{f}_\lambda \end{bmatrix} \begin{bmatrix} d\theta \\ dV \\ d\lambda \end{bmatrix} = 0 \tag{40}$$

The left hand side of the equation is a matrix of partial derivatives that multiplies the tangent vector form with differential elements. The matrix of partial derivatives is known as the Jacobian of the conventional load flow problem that is augmented by the column $\mathbf{F}_\lambda$, which can be obtained by taking the partial derivatives with respect to $\lambda$ (35) and (36), which gives:

$$\begin{bmatrix} \Delta P \\ \Delta Q \end{bmatrix} = \begin{bmatrix} H & N & -P^{esp} \\ J & L & -Q^{esp} \end{bmatrix} \begin{bmatrix} d\theta \\ \dfrac{dV}{V} \\ d\lambda \end{bmatrix} \tag{41}$$

Due to the nature of (41) which is a set of $nc+n-1$ equations with $nc+n$ unknowns and by adding $\lambda$ to the load flow equations, it is not possible to find a unique and nontrivial solution of the tangent vector; consequently an additional equation is needed.

This problem is solved by selecting a magnitude different from zero for one of the components of the tangent vector. In other words, if the tangent vector is denoted by:

$$\mathbf{t} = \begin{bmatrix} d\theta \\ dV \\ d\lambda \end{bmatrix} = 0, \quad t_k = \pm 1 \tag{42}$$

which leads to:

$$\begin{bmatrix} \mathbf{H} & \mathbf{N} & -\mathbf{P}^{esp} \\ \mathbf{J} & \mathbf{L} & -\mathbf{Q}^{esp} \\ & \mathbf{e}_k & \end{bmatrix} \cdot \mathbf{t} = \begin{bmatrix} \mathbf{0} \\ \pm 1 \end{bmatrix} \tag{43}$$

where $\mathbf{e}_k$ is a vector of dimension m+1 with all elements equal to zero but the $k$-th one, which is equal to 1. If the index $k$ is correctly selected, $t_k = \pm 1$ impose a none zero norm to the tangent vector and it guarantees that the augmented Jacobian will be nonsingular at the critical point (Eheinboldt et al, 1986). The usage of +1 or -1 depends on how the $k$-th state variable is changing during the trajectory solution which is being plotted. In a next section of this chapter, a method to select $k$ will be presented. Once the tangent vector has found the solution of (43), the prediction is carried out as follows:

$$\begin{bmatrix} \mathbf{\theta}^* \\ \mathbf{V}^* \\ \lambda^* \end{bmatrix} = \begin{bmatrix} \mathbf{\theta} \\ \mathbf{V} \\ \lambda \end{bmatrix} + \sigma \begin{bmatrix} d\mathbf{\theta} \\ d\mathbf{V} \\ d\lambda \end{bmatrix} \tag{44}$$

where "*" denotes the prediction for a future value of $\lambda$ (the load parameter) and $\sigma$ is a scalar that defines the step size. One inconvenient in the control of the step size consists in its dependence of the normalized tangent vector (Alves et al, 2000),

$$\sigma = \frac{\sigma_0}{\|\mathbf{t}\|} \tag{45}$$

where $|t_k|$ is the Euclidian norm of the tangent vector and $\sigma_0$ is a predefined scalar. The process efficiency depends on making a good selection of $\sigma_0$, which its value is system dependant.

### 4.5.2 Parameterization and corrector
After the prediction has been made, it is necessary to correct the approximate solution. Every continuation technique has a particular parameterization that gives a way to identify the solution along the plotting trajectory. The scheme here presented is referred as a local parameterization. In this scheme, the original set of equations is augmented with one extra equation, which has a meaning of specifying the value of a single state variable. In the case of the reformulated equations, this has a meaning of giving a unity magnitude to each nodal voltage, the phase angle of the nodal voltage or the load parameter $\lambda$. The new set of equations involves the new definition of state variables as:

$$\mathbf{x} = \begin{bmatrix} \mathbf{\theta} \\ \mathbf{V} \\ \lambda \end{bmatrix} \tag{46}$$

where,

$$x_k = \eta \tag{47}$$

where $x_k \varepsilon \mathbf{x}$ and $\eta$ represent the appropriate $k$-th element of $\mathbf{x}$. Therefore, the new set of equations that substitute (38) is given by,

$$\begin{bmatrix} \mathbf{F(x)} \\ x_k - \eta \end{bmatrix} = 0 \tag{48}$$

After an appropriate index $k$ has been selected and its corresponding value of $\eta$, the load flow is solved with the slightly modified Newton Raphson method (48). In other words, the $k$ index used in the corrector is the same as that used in the predictor and $\eta$ is equal to the obtained $x_k$ from the corrector (44), thus the variable $x_k$ is the continuation parameter.
The application of the Newton-Raphson to (38) results in,

$$\begin{bmatrix} H & N & -\mathbf{P}^{esp} \\ M & L & -\mathbf{Q}^{esp} \\ & \mathbf{e}_k & \end{bmatrix} \begin{bmatrix} \Delta\theta \\ \Delta\mathbf{V} \\ \Delta\lambda \end{bmatrix} = \begin{bmatrix} \Delta\mathbf{P} \\ \Delta\mathbf{Q} \\ 0 \end{bmatrix} \tag{49}$$

where $\mathbf{e}_k$ is the same vector used in (43), and the elements $\Delta\mathbf{P}$ and $\Delta\mathbf{Q}$ are calculated from (35) and (36). Once the $x_k$ is specified in (48), the values of the other variables are dependant on it and they are solved by the iterative application of (49).

### 4.5.3 Selection of the continuation parameter

The most appropriate selection corresponds to that state variable with the component of the tangent vector with the largest rate of change relative to the given solution. Typically, the load parameter is the best starting selection, i.e. $\lambda=1$. This is true if the starting base case is characterized for a light or normal load; in such conditions, the magnitudes of the nodal voltages and angles keep almost constants with load changes. On the other hand, when the load parameter has been increased for a given number of continuation steps, the solution trajectory is approximated to the critical point, and the voltage magnitudes and angles probably will have more significant changes. At this point, $\lambda$ has had a poor selection as compared with other state variables. Then, once the first step selection has been made, the following verification must be made:

$$x_k \leftarrow \max \left\{ \frac{t_1}{x_1} \quad \frac{t_2}{x_2} \quad \dots \quad \frac{t_{m-1}}{xm-1} \quad \frac{t_m}{x_m} \right\} \tag{50}$$

where $m$ is equal to the state variables; including the load parameter and $k$ corresponding to the maximum $t/x$ component. When the continuation parameter is selected, the sign of the corresponding component of the tangent vector must be taken into account to assign +1 or -1 to $t_k$ in (42) for the subsequent calculation of the tangent vector.

## 5. MATLAB resources for electrical networks

The reason why MATLAB is frequently chosen for the development of academic or research tools is because its huge amount of mathematical operations as those related to vectors and matrixes. In addition, it also has several specialized libraries (toolboxes) for more specific areas as control, optimization, symbolic mathematics, etc. In the area of power systems, it is possible to point it out several advantages considered as key points for the development of a script program, which are discussed below.

## 5.1 Sparse matrix manipulation

The electrical networks are studied using nodal analysis, as it was presented in section 3.1. In this frame of reference, the network matrixes as Ybus or Jacobian (1) and (21) have a sparse structure, considering that a node in an electrical network is connected to about 2.4 nodes in average. To illustrate this issue, if we consider the creation of a squared matrix with order 20, i.e. the matrix has 200 elements which 256 are zeros and the remaining are different from zero that are denoted as nz, as it is shown in figure 4. The sparsity pattern is displayed by using the MATLAB function *spy*.

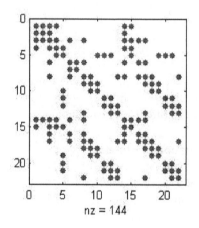

Fig. 4. An example of a sparse matrix.

The operation on this type of matrices with conventional computational methods leads to obtain prohibited calculation times (Gracia et al, 2005). Due to this reason, there have been adopted special techniques for deal with this type of matrixes to avoid the unnecessary usage of memory and to execute the calculation processes on the nonzero elements. It is important to point it out that this kind of matrixes is more related to a computation technique than to a mathematic concept.

MATLAB offers mechanisms and methods to create, manipulate and operate on this kind of matrixes. The sparse matrixes are created with the sparse function, which requires the specification of 5 arguments in the following order: 3 arrays to specify the position $i$, the position $j$ – row and column– and the element values x that correspond to each position (i,j), and the two integer variables to determine the dimensions *mxn* of the matrix, for example:

>> A = sparse(i, j, x, m, n)

In MATLAB language, the indexing of full matrixes is equal to the sparse matrixes. This mechanism consists in pointing to a set of matrix elements through the use of two arrays that makes reference to each row and column of the matrix, e.g. **B** = **A[I, J]** or by simply pointing to each element o elements of the matrix to be modified, e.g. A[**I, J**] = **X.**

where **A** is a *mxn* spare matrix, I and J are the arrays that point to the rows and columns and **X** is the array that contains each element corresponding to each ordered pair ($I_k$, $J_k$). All mentioned arrays are composed for a number of $k$ elements, such as $k < n$ and $k < m$.

MATLAB has a function called spdiags that is intended for the direct operation on any diagonal of the matrix, e.g. to make uniform changes on the elements of the diagonal "d" of matrix A:

>> A = spdiags(B, d, A)

A mechanism frequently used to form matrixes from other defined ones is called the *concatenation*. Even more, to form sparse matrixes by using other arrays of the same type but with specified dimensions in such a way that there is a consistency to gather into a single one, i.e. the concatenating is horizontal, the matrix rows must be equal to those of B matrix, e.g. C = [A B]. On the other hand, if the concatenating is vertical, then the columns of both matrices must be equal, e.g. C = [A; B].

### 5.2 The LU factorization for solving a set of equations

In electrical network applications, to find a solution of the algebraic system $\mathbf{Ax} = \mathbf{b}$ in an efficient way, one option is to use the triangular decomposition as the LU factorization technique.

The triangular factorization LU consists in decomposing a matrix (A) such that it can be represented as the product of two matrixes, one of them is a lower triangular (L) while the other is an upper triangular (U), $\mathbf{A} = \mathbf{L} \cdot \mathbf{U}$. This representation is commonly named explicit factorization LU; even though it is very related to the Gaussian elimination, the elements of L and U are directly calculated from the $\mathbf{A}$ elements, the principal advantage with respect to the Gaussian elimination (Gill et al., 1991) consists in obtaining the solution of an algebraic system for any $\mathbf{b}$ vector, if and only if the "$\mathbf{A}$" matrix is not modified.

### 5.3 Sparse matrix ordering using AMD

The Approximate minimum degree permutation (AMD) is a set of subroutines for row and column permutation of a sparse matrix before executing the Cholesky factorization or for the LU factorization with a diagonal pivoting (Tim, 2004). The employment of this subroutine is made by using the function "amd" to the matrix to be permuted, e.g. P = amd(**A**).

### 5.4 Sparse matrix operations

Care must be taken with a several rules in MATLAB when operations are carried out that include full and sparse matrixes, for example, the operation eye(22) + speye(22) gives a full matrix.

### 5.5 Vector operations

An approach to develop the script programs in MATLAB to be executed faster consists in coding the algorithms with the use of vectors within the programs and avoiding the use of loops such as *for, while and do-while*. The vector operations are made by writing the symbol "." before the operation to be made, e.g. .+, .-, .*, ./. This discussion is illustrated by comparing two MATLAB script programs to solve the following operation:

$$\begin{bmatrix} C_1 \\ C_2 \\ \vdots \\ C_n \end{bmatrix} = \begin{bmatrix} A_1 \cdot B_1 \\ A_1 \cdot B_2 \\ \vdots \\ A_n \cdot B_n \end{bmatrix}$$

| Script program with the loop command | Script program with vector operations. |
|---|---|
| for k = 1:1000<br>    C(k,1) = k*k;<br>end<br>*Execution time*: 0.002086 s | A = 1:1000;<br>B = 1:1000;<br>C = A.*B;<br>*Execution time*: 0.000045 s |

It is evident that the use of the *for-loop* command to carry out the described operation gives a larger computation time than by using the vector operation **A.*B**.

## 6. MATLAB application for plotting the PV curves

The program for plotting the PV curves is integrated with four specific tasks:
- Reading of input data
- Generation of the flow base case
- Calculation of points for the PV curves at each node.
- Graphical interface for editing the display of PV curves.

Afterwards, the complete application code is listed. In case the code is copied and pasted into a new m-file, this will have the complete application, i.e. there is no necessity to add new code lines.

```
function main()
%% *************    READING OF INPUT DATA    *************

% -> The filename and its path are obtained:
  [file, trayectoria] = uigetfile( ...
    { '*.cdf;','Type file (*.cdf)'; '*.cf', ...
    'IEEE Common Format (*.cf)'}, 'Select any load flow file');

  if file == 0 & trayectoria == 0
    return;
  end
  archivo = [trayectoria file];

% -> the code lines are stored in the following variable:
  DATA = textread(archivo,'%s','delimiter','\n','whitespace','');

% -> Reading of nodal data:
  X_b = DATA{1};
  Sbase = str2double(X_b(32:37)); % Base power (MVA): Sbase

  X_b = DATA{2};
  N_Bus = findstr(X_b,'ITEMS');
  N_Bus = str2double(fliplr(strtok(fliplr(X_b(1:N_Bus-1)))));
  Nnod  = N_Bus;  % Total number of buses
  N_Bus = 2+N_Bus;
  DN   = strvcat(DATA{3:N_Bus});
```

```
No_B = str2num(DN(:,1:4));      % Number of buses: No_B
Tipo = str2num(DN(:,25:26));    % Bus type: Tipo

Nslack = find(Tipo == 3);  % Slack bus
NumNG  = find(Tipo >= 2);  % Index of PV nodes
NumNC  = find(Tipo == 0);  % Index of PQ nodes
NumCG  = find(Tipo <= 2);  % Index of all nodes except slack

Vabs = str2num(DN(:,28:33));    % Magnitude of nodal voltages.
Tetan = str2num(DN(:,34:40));   % Phase angle of nodal voltages (degrees)
Tetan = Tetan*pi/180;           % Degrees to radians conversion

Pload = str2num(DN(:,41:49));   % Demanded active power (MW)
Qload = str2num(DN(:,50:59));   % Demanded reactive power (MVAr)

Pgen = str2num(DN(:,60:67));    % Generated active power (MW)
Qgen = str2num(DN(:,68:75));    % Generated reactive power (MVAr)
Pgen = Pgen(NumNG);     % PV nodes are selected
Qgen = Qgen(NumNG);     %PV nodes are selected
Gc   = str2num(DN(:,107:114)); % Compensator conductance
Bc   = str2num(DN(:,115:122)); % Compensator susceptance
NumND = find(Bc~=0);    % Indexes of all compensating nodes.
Bcomp = Bc(NumND);      % Susceptance of all compensating nodes  NumND

% -> Branch data is read:
 X_b = DATA{N_Bus+2};
 N_Bra = findstr(X_b,'ITEMS');
 N_Bra = str2double(fliplr(strtok(fliplr(X_b(1:N_Bra-1)))));
 N_Bra1 = N_Bus+3;
 N_Bra2 = N_Bus+N_Bra+2;
 XLN = strvcat(DATA{N_Bra1:N_Bra2});

 X_b   = str2num(XLN(:,19));
 NramL = find(X_b == 0);   % Indexes of transmission lines (LT)
 NramT = find(X_b >= 1);   % Indexes of transformers (Trafos)
 PLi   = str2num(XLN(NramL,1:4));  % Initial bus for the transmission lines
 QLi   = str2num(XLN(NramL,6:9));  % Final bus for the transmission lines
 PTr   = str2num(XLN(NramT,1:4));  % Initial bus for the transformers
 QTr   = str2num(XLN(NramT,6:9));  % Final bus for the transformers

 RbrL = str2num(XLN(NramL,20:29)); % Resistance of transmission lines (pu)
 XbrL = str2num(XLN(NramL,30:40)); % Reactance of transmission lines (pu)
 Blin = str2num(XLN(NramL,41:50))/2; % Susceptance of transmission lines (pu)

 RbrX = str2num(XLN(NramT,20:29)); % Resistance of transmission lines (pu)
 XbrX = str2num(XLN(NramT,30:40)); % Reactance of transmission lines (pu)
```

```
    TAP  = str2num(XLN(NramT,77:82));   % TAP changers of transformers

% -> Admittance matrix:

    Ylin =1./(RbrL+1j*XbrL);        % Series admítanse of transmission lines
    Yxtr =(1./(RbrX+1j*XbrX))./TAP; % Series admítanse of transformers
%   Mutual admittances:
    Ynodo = sparse([ PLi;  PTr;  QLi;  QTr ],...
              [ QLi;  QTr;  PLi;  PTr ],...
              [-Ylin; -Yxtr; -Ylin; -Yxtr ], Nnod, Nnod);

%  Self admittances:
   Ynodo = Ynodo + ...
   sparse([      PLi;      QLi;     PTr;     QTr;   NumND],...
     [      PLi;      QLi;     PTr;     QTr;   NumND],...
     [Ylin+1j*Blin; Ylin+1j*Blin; Yxtr./TAP; Yxtr.*TAP; 1j*Bcomp],...
     Nnod, Nnod);

%% ******************  CASE BASE GENERATION   ********************

   Tol_NR  = 1e-3;   % Tolerance error                    .
   Max_Iter = 60;      % Maximum number of possible iterations.

%  -> Initialization:
   Vabs(NumNC) = 1.0;  % Voltage magnitude of PQ nodes
   Tetan      = zeros(Nnod,1);   % Phase angles in all nodes

%  -> Calculation of specified powers:
%  Active power:
   PLg = Pload(NumNG);
   QLg = Qload(NumNG);
   Pesp(NumNG,1)  = (Pgen-PLg)/Sbase;     % for PV nodes
   Pesp(NumNC,1)  = (-Pload(NumNC))/Sbase; % for PQ nodes
%  Reactive power:
   Qesp  = -Qload/Sbase;  % for PQ nodes

   Ndim = Nnod + length(NumNC) - 1;   % Jacobian dimension.
   iter = 0;  % Initialization of iteration counter.
   tic

   while iter <= Max_Iter  % Limit of the number of iterations
%  -> Calculation of net active and reactive powers injected to each node:
     [Pnodo Qnodo] = scmplx();

%  -> Calculation of  misadjustment of reactive and active powers:
```

```
        DelP  = Pesp(NumCG) - Pnodo(NumCG);
        DelQ  = Qesp(NumNC) - Qnodo(NumNC);
        Pmismatch = [DelP; DelQ];

%  -> Check the convergence method.
        if max(abs(Pmismatch)) < Tol_NR
            tsol = toc;
            fprintf( 'The case has converged: %d iterations \n', iter);
            fprintf( 'in: %f seg. \n', tsol);
            break;
        end

%  -> Jacobian calculation:
        JB = Jacob();

%  -> Solution of the set of equations:
        DeltaX = linear_solver(JB, Pmismatch);

%  -> Update of magnitudes and angles of nodal voltages:
        Vabs(NumNC) = Vabs(NumNC).*(1 + DeltaX(Nnod:Ndim));
        Tetan(NumCG) = Tetan(NumCG) + DeltaX(1:Nnod-1);

        iter = iter + 1;
    end

%% ******   CALCULATION OF POINTS OF PV CURVES AT EACH NODE   ******

    Ndim = Nnod + length(NumNC); % Augmented Jacobian dimension.

    sigma0   = 0.3966;    % Predictor step.
    Vpredict = [Vabs];    % Predicted voltage values.
    Vexact   = [Vabs];    % Corrected voltage values.
    Carga    = [0];       % Load increment.
    lambda   = 0;         % Initial value of lambda.
    b(Ndim,1) = 1;        % Column vector of equation (43).
    Ix = Ndim;            % Index that indicates the continuation parameter.

% Flag to indicate state before the critical point:
    Superior  = true;

% Counter used to calculate exactly the number of points alter the critical point:
    Inferior  = 0;

    while(Superior == true) || (Inferior > 0)
%  -> After the critical point has been reached, only more 14 points are obtained.
        if (Superior == false) Inferior = Inferior - 1; end
```

```
%   -> Calculation of the augmented Jacobian as in equation (43):
    JB = Jacob();
    JB(Ndim,Ix) = 1;
    JB(1:Ndim-1,Ndim) = [-Pesp(NumCG); -Qesp(NumNC)];

%   -> Calculation of the tangent vector as in equation (43)
    t = linear_solver(JB, b);

%   -> The critical point has been reached:
    if b(end) < 0
      if Superior == true;
        sigma0  = 0.25;   % Decrement of the predictor step
        Inferior = 14;      % Calculation of additional 14 points.
        Superior = false;

        Carga(end-1:end)     = [];
        Vpredict(:, end-1:end) = [];
        Vexact(:, end)        = [];
      end
    end

%   -> Selection of the continuation parameter:
    x  = [Tetan(NumCG); Vabs(NumNC); lambda];
    [xk Ix] = max(abs(t)./x);   % As in equation (50)

%   -> Calculation of the predictor step as in equation (45):
    sigma = sigma0/norm(t);

%   -> Calculation of predictor:
    Vabs(NumNC)  = Vabs(NumNC).*(1 + t(Nnod:Ndim-1)*sigma);
    Tetan(NumCG) = Tetan(NumCG) + t(1:Nnod-1)*sigma;
    lambda = lambda + sigma*t(end);

    Vpredict = [Vpredict Vabs]; % Calculated values with the predictor
    Carga  = [Carga lambda];    % Load increment

%   -> Calculation with the corrector:
    iter = 0;
    while iter <= Max_Iter

      [Pnodo Qnodo] = scmplx();   % Computation of the complex power
%        -> Calculation of misadjustment  in active and reactive powers as in:
        DelP = Pesp(NumCG)*(1 + lambda) - Pnodo(NumCG); % Equation (35)
        DelQ = Qesp(NumNC)*(1 + lambda) - Qnodo(NumNC); % Equation (36)
        Pmismatch = [DelP; DelQ; 0];
```

```
        if max(abs(Pmismatch)) < Tol_NR    % Convergence criterion
            break;
        end
%       -> Calculation of the augmented Jacobian.
        JB = Jacob();
        JB(1:Ndim-1,Ndim) = [-Pesp(NumCG); -Qesp(NumNC)];
        JB(Ndim,Ix) = 1;

%       -> Calculation of the set of equations as in (49).
        DeltaX = linear_solver(JB, Pmismatch);

%       -> Checking if the critical point has been found.
        if DeltaX(end) < 0 && Superior == true
            b(Ndim,1) = -1;
            break;
        end

%       -> Update state variables in the corrector:
        Vabs(NumNC)  = Vabs(NumNC).*(1 + DeltaX(Nnod:Ndim-1));
        Tetan(NumCG) = Tetan(NumCG) + DeltaX(1:Nnod-1);
        lambda       = lambda + DeltaX(end);

        iter = iter + 1;
    end

    Vexact   = [Vexact Vabs];
    Vpredict = [Vpredict Vabs];
    Carga    = [Carga lambda];

end

%% **   GRAPHICAL INTERFACE FOR DISPLAYING THE PV CURVES   **
PV_PRINT( Carga, Vpredict, Vexact );

%% ********************** Nested functions ***********************
function [P Q] = scmplx()

    Vfasor = Vabs.*exp(1j*Tetan);      % Complex form of nodal voltage using Euler.
    Scal = Vfasor.*conj(Ynodo*Vfasor);   % Calculation of complex power as in (8).
    P = real(Scal); % Net active power injected to each node as in (9).
    Q = imag(Scal); % Net reactive power injected to each node as in (10).

end

    function Jac = Jacob()
```

```
    Jac = sparse(Ndim, Ndim);

    Vdiag = sparse(1:Nnod, 1:Nnod, Vabs.*exp(1j*Tetan));
    J2   = Vdiag*conj(Ynodo*Vdiag);   % Computation of equation (34)
    P = sparse(1:Nnod, 1:Nnod, Pnodo);
    Q = sparse(1:Nnod, 1:Nnod, Qnodo);

    H = ( imag(J2(NumCG, NumCG)) - Q(NumCG, NumCG));   % Equations  (26) and (30)
    N = ( real(J2(NumCG, NumNC)) + P(NumCG, NumNC));   % Equations (27) and (31)
    J = (-real(J2(NumNC, NumCG)) + P(NumNC, NumCG));   % Equations (28) and (32)
    L = ( imag(J2(NumNC, NumNC)) + Q(NumNC, NumNC));   % Equations (29) and (33)

    Jac = horzcat( vertcat(H, J), vertcat(N, L) );  % Submatrix concatenation as indicated in
(25)

  end

  function x = linear_solver(A, b)

    P = amd(A);         % Ordering of matrix A (Jacobian).
    [L, U]  = lu(A(P, P));  % Application of LU factorization.
  % The set of equations (b=A*x) are solved using the LU factorization:
    y     = L\b(P);
    x(P, 1) = U\y;

  end

end
```

As it was shown above, the four specific tasks of this application are integrated around the principal function: main(), which function is the sequential integration of these tasks. Even though the first three tasks are described in a set of continuous code lines, each task represents one independent process from each other; hence it is possible to separate them as external functions, such as the function PV_PRINT, which executes the four tasks. Additionally, it can be seen that the calculation tasks, flow case base and the points of PV curves use nested functions within the main function that are common to both tasks. The following nested functions are:

*Scmplx*: This function computes the net complex power injected to each node of the system and it returns its real and imaginary components.

*Jacob*: This function computes the Jacobian of the conventional load flow problem.

*Linear_solver*: This function solves the set of linear equations $\mathbf{A} \cdot \mathbf{x} = \mathbf{b}$ by using the LU factorization with previous reordering of $\mathbf{A}$ matrix.

### 6.1 Reading of input data
A program based on the solution of load flows requires of some input data that describes the network to be analyzed. The information can be included in a text file with a

standardized format with the extension cdf (Common Data Format) (University of Washington electrical engineering, 1999). The reading of the complete information is achieved by using the MATLAB function textread. The reading includes the use of a standard uicontrol of MATLAB to locate the file path. This information is stored in a data matrix of string type, and it can be interpreted by using functions that operate on the variables of string type. There is basic information that is used in the developed program; the variables can be divided into two categories:

*Nodal information:*

*Sbase* : Base power of the system.

*Nnod* : Number of system nodes.

**No_B** : $n$x1 vector with the numbering of each system bus.

**Name_B** : $n$x1 vector with the name of each bus system.

**Tipo** : $n$x1 vector with the node type of the system.

*Nslack* : The number of the slack node.

**Vn** : $n$x1 vector with voltage magnitudes at each system node.

**Vbase** : $n$x1 vector with base values of voltages at each system bus.

**Tetan** : $n$x1 vector with phase angles at each nodal voltage.

**PLoad** and **QLoad** : $n$x1 vector with the active and reactive load powers, respectively.

**Pgen** and **Qgen** : $(n-nc)$x1 vector with the active and reactive power of generators, respectively.

**Gc** and **Bc** : $n$x1 vector with conductance and susceptance of each compensating element, respectively.

*Branch information* (transmission lines and transformers), where $l$ denotes the total number of transmission lines and $t$ is the number of transformers in the electrical network:

**PLi** and **PTr** : Vectors with dimensions $l$x1 y $t$x1, that indicate one side of transmission lines and transformers connected to the P node, respectively.

**QLr** and **QTr** : They are similar as PL and PT with the difference that indicates the opposite side of the connected branch to the Q node.

**RbrL** and **RbrX** : They are similar as the above two variables with the difference that indicate the resistance of each transmission line and transformer.

**XbrL** and **XbrX** : They are similar as the above with the difference that indicate the reactance of each transmission line and transformer.

**Blin** : $l$x1 vector with the half susceptance of a transmission line.

**TAP** : $t$x1 vector with the current position of the TAP changer of the transformer.

After reading the input data, the admittance matrix **Ynode**, and a series of useful pointers are created to extract particular data from the nodal and branch data:

- **NumNG**: An array that points to the PV nodes.
- **NumNC**: An array that points to the PQ nodes.
- **NumCG**: An array that points to all nodes but the slack node.

*Tol_NR – the maximum deviations in power are lower to this value in case of convergence - and Max_Iter –is the maximum number of iterations –* these variables are used to control the convergence and a value is assigned to them by default after reading of input data.

The program does not consider the checking of limits of reactive power in generator nodes neither the effect of the automatic tap changer in transformers, nor the inter-area power transfer, etc. Therefore the input data related to these controllers is not used in the program.

## 6.2 Graphical interface for editing the display of PV curves

In the last part of the program, a graphical user interface (GUI) is generated for the plotting of PV curves from the points previously calculated in the continuous load flows program. Even though the main objective of this interface is oriented to the plotting of PV curves, its design allows the illustration continuation method process, i.e. it has the option of displaying the calculated points with the predictor, corrector or both of them, for example, it can be seen in figure 5 the set of points obtained with the predictor-corrector for the nodes 10,12,13, and 14 of the 14-node IEEE test power system. This is possible by integrating the MATLAB controls called uicontrol´s into the GUI design, the resultant GUI makes more intuitive and flexible its use.

The numbers 1-12, that are indicated in red color in figure 5, make reference to each uicontrol or graphical object that integrates the GUI. The graphical objects with its uicontrol number, its handle and its description are given in table 1.

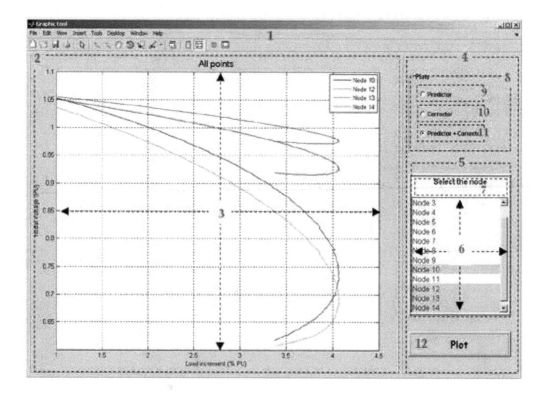

Fig. 5. GUI for plotting the PV curves for any node.

| Uicontrol number | handle | Description |
|---|---|---|
| 1 | hPV | Graphical object known as figure |
| 2 | hPanel_print | Panel located to the left of hPV |
| 3 | hAxes_Curve | Object for 2D graphs known as axes |
| 4 | hPanel_control | Panel located to the right of hPV |
| 5 | hPanel | Panel located in the upper part of hPanel_control |
| 6 | hlb_log | Graphical object known as listbox |
| 7 | hlb_txt | Graphical object known as textbox |
| 8 | hChoice_PV | Set of objects known as radio buttons |
| 9 | opt0 | Graphical objects known as radiobutton |
| 10 | opt1 | Graphical objects known as radiobutton |
| 11 | opt2 | Graphical objects known as radiobutton |
| 12 | - without handle- | Graphical objects known as pushbutton |

Table 1. Description of each graphical object used in the GUI shown in figure 5.

The complete code of the function that generates the GUI is presented below, where the code lines for creating each graphical object mentioned in table 1 are fully described.

```
function PV_PRINT(lambda, predictor, corrector )

   [Nodos columnas] = size(predictor);
   clear columnas;
   etiqueta = {};


% Variable for registering the elements of listbox that relates the system nodes, which can
have plotted its PV curve
   for ii = 1:Nodos
      etiqueta{ii, 1} = ['Node ' num2str(ii)];
   end


% Variables for configuring the graphical object, color and screen size.
   BckgrClr = get(0, 'defaultUicontrolbackgroundColor');
   Sc_Sz = get(0, 'ScreenSize');
   wF = 0.85;
   hF = 0.85;
   wF_Pix = wF*Sc_Sz(3);
   hF_Pix = hF*Sc_Sz(4);


% Creation of the object type figure: main window.
   hPV = figure(...
         'Name', 'Graphic tool', ...
         'units', 'normalized', 'Color', BckgrClr,...
```

```
        'MenuBar', 'figure', 'numbertitle', 'off',...
        'visible', 'on', 'Position', [0.05 0.05 wF hF]);
    delete(gca);

% Generation of the left panel included in hPV.
    hPanel_print = uipanel('Parent', hPV, ...
        'Units', 'normalized', 'BackgroundColor', BckgrClr, ...
        'BorderType', 'etchedout', 'BorderWidth', 1,...
        'visible', 'on', 'Position', [0.005 0.01 0.75 0.98]);

% Declaration of axes object where the curves are to be plotted and it is located in
hPanel_Print.
    hAxes_Curve = axes('Parent', hPanel_print, ...
        'units', 'normalized', 'Position',[0.07 0.07 0.88 0.88], ...
        'Box', 'on', 'Tag', 'hAxes_Curve', 'Visible', 'off');

% Generation of the right panel in hPV.
    hPanel_control = uipanel('Parent',hPV, ...
        'Units', 'normalized', 'BackgroundColor', BckgrClr, ...
        'BorderType', 'etchedout', 'BorderWidth', 1,...
        'visible', 'on', 'Position', [0.76 0.01 0.235 0.98]);

% Declaration of a panel inside the hPanel_control  and it contains the listbox and textbox.
    hPanel = uipanel('Parent', hPanel_control, ...
        'Units', 'normalized', 'BackgroundColor', BckgrClr, ...
        'BorderType', 'etchedout', 'BorderWidth', 3,...
        'visible', 'on', 'Position', [0.05 0.18 0.9 0.45]);

% Listbox located in hPanel and it is used for selecting the node or desired nodes for
displaying its curves.
    hlb_log = uicontrol('Parent', hPanel, ...
        'Units', 'normalized', 'BackgroundColor', [1 1 1], ...
        'FontName', 'Lucida Console', 'HorizontalAlignment', 'left',...
        'Tag', 'hlb_log', 'Position', [0 0 1 0.85], 'fontsize', 12,...
        'Enable', 'on', 'Style', 'listbox', 'String', etiqueta, ...
        'Max', Nodos, 'Min', 0, 'Clipping', 'off', 'Visible', 'on');

% Textbox located in hPanel and it is used for displaying any string.
    hlb_txt = uicontrol('Parent', hPanel,...
        'Units', 'normalized', 'HorizontalAlignment', 'center',...
        'style', 'text', 'fontname', 'Comic Sans MS',...
        'BackgroundColor', 'w', 'string', 'Select the node',...
        'fontsize', 12, 'fontweight', 'bold', 'visible', 'on',...
        'Position', [0.0 0.85 1 0.15]);

% Uicontrol of type optionbutton, and it is located in hPanel_control.
    hChoice_PV = uibuttongroup('Parent', hPanel_control, ...
```

```
       'Units', 'normalized', 'Position', [0.05 0.7 0.9 0.25], ...
       'BackgroundColor', BckgrClr, 'FontName', 'Lucida Console', ...
       'FontSize', 10, 'FontWeight', 'bold', 'Tag','hChoice_PV',...
       'Title', 'Plots', 'visible', 'off');
```

% Optionbutton located in hChoice_PV which is used for ploting the points obtained with
the corrector.
```
   opt0 = uicontrol('Parent', hChoice_PV, ...
       'Units', 'normalized', 'pos', [0.1 0.68 0.8 0.25], ...
       'BackgroundColor', BckgrClr, 'FontName', 'Lucida Console', ...
       'FontSize', 9, 'FontWeight', 'bold', 'Style', 'Radio',...
       'String', 'Predictor', 'HandleVisibility', 'off');
```

% Optionbutton located in hChoice_PV and it is used for plotting the points calculated with
the corrector.
```
   opt1 = uicontrol('Parent', hChoice_PV, ...
       'Units', 'normalized', 'pos', [0.1 0.4 0.8 0.25], ...
       'BackgroundColor', BckgrClr, 'FontName', 'Lucida Console', ...
       'FontSize', 9, 'FontWeight', 'bold', 'Style', 'Radio',...
       'String', 'Corrector', 'HandleVisibility', 'off');
```
% Optionbutton located in hChoice_PV  and it is used for plotting the points obtained with
the predictor + corrector.
```
   opt2 = uicontrol('Parent', hChoice_PV, ...
       'Units', 'normalized', 'pos', [0.1 0.125 0.8 0.25], ...
       'BackgroundColor', BckgrClr, 'FontName', 'Lucida Console', ...
       'FontSize', 9, 'FontWeight', 'bold', 'Style', 'Radio',...
       'String', 'Predictor + Corrector', 'HandleVisibility', 'off');
```

% Configuration of the graphical object hChoice_PV.
```
   set(hChoice_PV, 'SelectionChangeFcn', @selcbk);
   set(hChoice_PV, 'SelectedObject', opt0);
   set(hChoice_PV, 'UserData', 0);
   set(hChoice_PV, 'Visible', 'on');
```

% Generation of the command button which calls the function that generates the selected
graphs for the set of optionbutton located in hChoice_PV.
```
   uicontrol('Parent', hPanel_control, ...
       'Units', 'normalized', 'style', 'pushbutton',...
       'FontName', 'Comic Sans MS', 'visible', 'on',...
       'fontsize', 9, ...
       'callback', {@RefreshAxes, hChoice_PV, hAxes_Curve, hlb_log, lambda, predictor,
corrector},...
       'FontWeight', 'bold', 'string', 'Plot', ...
       'Position',[0.05 0.05 0.9 0.08]);
```

end

```
function selcbk(source, eventdata)

    switch get(eventdata.NewValue, 'String')
      case 'Predictor'
        set(source, 'UserData', 0);
      case 'Corrector'
        set(source, 'UserData', 1);
      otherwise
        set(source, 'UserData', 2);

    end

end

function RefreshAxes(source, eventdata, hChoice_PV, hAxes_Curve, hlb_log, lambda,
predictor, corrector)

  Nodos = get(hlb_log, 'Value');      % It is obtained the node(s) whose PV curves will be
plotted
    switch get(hChoice_PV, 'UserData') % Option selection for the object uibuttongroup

      case 0  % The points obtained with the predictor are plotted for the selected nodes.
          plot(hAxes_Curve, 1 + [lambda(1), lambda(2:2:end)], [predictor(Nodos, 1),
predictor(Nodos, 2:2:end)]);
          title(hAxes_Curve, 'Plot the points of predictor', 'FontSize', 14);

        case 1  % The points obtained with the corrector are plotted for the selected nodes.
          plot(hAxes_Curve, 1 + lambda(1:2:end), corrector(Nodos,:));
          title(hAxes_Curve, 'Plot the points of corrector ', 'FontSize', 14);
        otherwise
        % The points stored in the variable predictor are plotted for the selected nodes.
          plot(hAxes_Curve, 1 + lambda, predictor(Nodos, :));
          title(hAxes_Curve, ' All points ', 'FontSize', 14);

    end

    if length(Nodos) > 1
    % If there are more than one node, they are labeled for marking the shown curves
      etiqN = get(hlb_log, 'String');
      legend(hAxes_Curve, etiqN{Nodos});
    end

% The grid are shown in the plot.
  grid(hAxes_Curve, 'on');

% Automatic numbering of axes (selection of maximum and minimum limits for x-y axes)
```

```
  axis(hAxes_Curve, 'auto');

% Labeling of X and Y axes.
  xlabel(hAxes_Curve, 'Load increment (% PU)', 'FontSize', 10);
  ylabel(hAxes_Curve, 'Nodal voltage (PU)', 'FontSize', 10);

% The object becomes visible with the handle hAxes_Curve.
  set(hAxes_Curve, 'Visible', 'on');

end
```

Since the function PV_PRINT is external to the function main, the created variables within main cannot be read directly for the function PV_PRINT, hence it is required to declare arguments as inputs for making reference to any variable in main. The main advantage of using external functions consists in the ability to call them in any application just by giving the required arguments for its correct performance. In order to reproduce this application, it is necessary to copy the code in an m-file which is different from the main.

## 7. Conclusion

An electrical engineering problem involves the solution of a series of formulations and mathematical algorithm definitions that describe the problem physics. The problems related to the control, operation and diagnostic of power systems as the steady-state security evaluation for the example the PV curves, are formulated in matrix form, which involves manipulation techniques and matrix operations; however the necessity of operating on matrixes with large dimensions takes us to look for computational tools for handling efficiently these large matrixes. The use of script programming which is oriented to scientific computing is currently widely used in the academic and research areas. By taking advantage of its mathematical features which are normally found in many science or engineering problems allows us solving any numerical problem. It can be adapted for the development of simulation programs and for illustrating the whole process in finding a solution to a defined problem, and thus makes easier to grasp the solution method, such as the conventional load flow problem solved with the Newton-Raphson method.

MATLAB has demonstrated to be a good tool for the numerical experimentation and for the study of engineering problems; it provides a set of functions that make simple and straightforward the programming. It also offers mechanisms that allow dealing with mathematic abstractions such as matrixes in such a way that is possible to develop prototype programs which are oriented to the solution methods by matrix computations. The development of scripts or tools can be considered to be a priority in the academic area such that they allow achieving a valid solution. It is also advisable to take advantage of the MATLAB resources such as: vector operations, functions and mechanisms for operating on each matrix element without using any flow control for the program, i.e. for loop; this offers the advantage of decreasing the number of code lines of the script program. These recommendations reduce the computation time and allow its easy usage and modification by any user. Finally, MATLAB offers powerful graphical tools which are extremely useful for displaying the output information and to aid interpreting the simulation results. In this chapter, the plotting of the corrector points (PV curve) has been presented for a given load that can be considered as critical point or voltage collapse of the power system. The technique for continuous load flow has been applied to determine the PV curves of the 14-node IEEE test system, and it has been shown that a 4.062% load increment can lead to the instability of the system, and it also has been determined that the node 14 is the weaker node of the system.

## 8. References

Arrillaga, J. & Walsion,N. (2001). *Computer Modeling of electrical Power Systems (2th)*, Whyley, ISBN 0-471-87249-0.

D. A. Alves, L. C. P. Da Silva, C. A. Castro, V. F. da Costa, "Continuation Load Flor Method Parametrized by Transmissión Line Powers", Paper 0-7801-6338-8/00 IEEE 2000

Garcia, J., Rodriguez, J.& Vidal, J. (2005). Aprenda Matlab 7.0 como si estuviera en primero, Universidad Politécnica de Madrid, Retrieved from http://mat21.etsii.upm.es/ayudainf/aprendainf/Matlab70/ matlab70primero.pdf

Gill, P., Murray, W. & Wrigth M., (1991). *Nuemric linear Algebra and optimization vol 1* (ed), Addison-wesley, ISBN 0-201-12647-4, Redwood City, US

Graham, R., Chow, J., &, Vanfretti, L. (2009). Power System Toolbox, In: Power System Toolbox Webpage, 15.03.20011, Available from: http://www.ecse.rpi.edu/pst/

Kundur, P. (1994). Voltage Stability, *Power System stability and control*, pp. 959-1024, Mc Graw Hill, ISBN 0-07-035959-X, Palo Alto, California

Mahseredjian, J., & Alvarado, F. (1997). MatEMTP, In: Creating an electromagnetic transients program in MATLAB, Available: 15.013.2011, http://minds.wisconsin.edu/handle/1793/9012

M. G. Ogrady, M. A. Pai, "Analysis of Voltage Collapse in Power Systems", Proceedings of 21st Annual North American Power Symposium, Rolla, Missouiri, October 1999ric Manufacturers Association, Washington, D.C.

Milano, F. (2010). *Power system modelling and scripting (1th)*, Springer, ISBN 9-783-642-13669-6, London, ___

Milano, F. (2006). PSAT, In: Dr. Federico Milano Website, 15.03.2011, http://www.power.uwaterloo.ca/~fmilano/psat.htm

PowerWorld Corporation (2010). http://www.powerworld.com

Siemens, PTI. (2005). http://www.energy.siemens.com/us/en/services/powertransmission-distribution/power-technologies-international/softwaresolutions/pss-e.htm

Tim, D. (2004), Research and software development in sparse matrix algorithms, In: AMD, 12.03.2011, Available from http://www.cise.ufl.edu/research/sparse/amd/

Venkataramana, A. (2007). *Power Computational Techniques for Voltage Stability Assessment and Control (1th)*, Springer, ISBN 978-0-387-26080-8.

Venkataramana Ajjarapu, Colin Christy, "The Continuation Power flor: A Tool for Steady State Voltaje Stability Analysis", Transactions on Power Systems, Vol. 7, No. 1, Fabruary 1992

W. C. Eheinboldt, J. B. Burkhardt, "A Locally Parametrized Continuation Process", ACM Transactions on Mathematical Software, Vol. 9 No. 2, June 1986, pp. 215-235.

University of Washington electrical engineering. (1999). 14 Bus, In: Power Systems Test Case Archive, 25.01.2011, Available from http://www.ee.washington.edu/research/pstca/.

Y. Yamura, K. Sakamoto, Y. Tayama, "Voltaje Instability Proximity Index baesd on Multiple Load Flor Solutions in Ill-Conditioned Power Systems". Proceedings of the 27th IEEE Conference on Decision and Control, Austin, Texas, December 1988

Zimmerman, R., Murillo, C., & Gan, D. (2011). MATPOWER, In: A MATLAB Power System Simulation Package, 15.03.2011, Available from: http://www.pserc.cornell.edu/matpower/

# 5

# A New Approach of Control System Design for LLC Resonant Converter

Peter Drgoňa, Michal Frivaldský and Anna Simonová
*University of Žilina, Faculty of Electrical Engineering, Žilina*
*Slovakia*

## 1. Introduction

Main task of power semiconductor converter is transfer of electrical energy from its input to the output and to the load. This energy flow is provided by control of the switching process of one or more semiconductor devices. Based on converter application it is necessary to choose correct control algorithm for selected switching frequency of converter. Algorithm itself has to be utilized with suitable hardware platform, whereby through the set of instructions the turn - on and turn - of control pulses are generated. Control system also has to secure other important functions for secondary processes. Nowadays a huge number of converter topologies exists, whereby each type should be suited for different conditions of use. For example uninterruptible power supply must generate constant output voltage with constant frequency and constant amplitude for wide range of output power. Converters for electric drive applications have to control their speed through generation of three phase voltage system with configurable amplitude and speed. From mentioned above it is clear to say that for each topology of power converter, the different type of control modulation (PWM - Pulse Width Modulation, PSM - Phase Shift Modulation, FM - Frequency Modulation) and different structure of regulator has to be used. Therefore selection of suitable control system depends on number of various factors, which are output power of converter, type of switching devices, type of load, possibility of hardware implementation and ways of switching.

In the field of DC/DC converters for switching power supplies, the main focus is on implementation of digital control, which means use of microcontrollers or DSP with implemented control algorithms and functions for communication with user as well. Analog control systems are well known and its design procedure is mastered, so it is easy to convert analog (continuous) controller into discrete and consequently implementing in microcontroller and DSP. But this approach ignore sample and hold effect of A/D converter and computing time of microprocessor. For proper discrete controller design, the method named direct digital design can be used. This method uses transfer function of converter in discrete form, including effect of sampling and computation delay.

For controller design purposes, transfer function of converter is necessary. This transfer function is based on model of power stage of controller. For PWM converters such as buck, boost, etc. the averaging method is widely used, but for new resonant topologies such as LLC converter, with this averaging method, the control to output transfer function cannot be used. In this chapter, new simulation based method for obtaining the transfer function of converter is discussed. This method uses MATLAB and OrCad PSPICE environments for revealing the transfer function. Advantage of this method is, that it can be used for every type of power converter. Next in the chapter, the direct digital design of controller based on transfer function is discussed. This method uses MATLAB environment for designing the discrete controller.

## 2. The ways of regulation of electric quantities in power converters

From the various types of regulation of electric quantities (voltage, current) the most well - known are PWM, PSM, and FM. In pulse width modulation (PWM) technique, the mean value of output voltage or current is controlled through the change of on-time ($t_{on}$) or off-time ($t_{off}$) of power transistor. Generation of PWM signals is based on comparison between repeating sequence (sawtooth signal or triangular signal with constant frequency) and between reference value of voltage ($u_{reg}$) which is generated from regulator. This principle is for analog and digital system almost the same. The only difference in the case of analog system is that repeating sequence is generated from op-amplifier and consequently compared with reference value in analog comparator. For generation of repeating sequence the digital system instead of op-amplifier utilizes digital timer. Comparator is replaced by compare register. The proportional on-time in the case of direct quantities is called duty cycle - D, and is defined as follows:

$$D = t_{on}/T \tag{1}$$

where $t_{on}$ is on-time of semiconductor devices, and $T_s$ is switching period of converter.

In the case of phase shift modulation (PSM) the mean value of output voltage is controlled by phase shift of on-times of two or more semiconductor devices, such two transistors in upper or lower leg of fullbridge converter are switched simultaneously, whereby switching frequency and duty cycle are constant. The application of PSM is well suited in ZVS (Zero Voltage Switching) and ZCS (Zero Current Switching) resonant converters.

Frequency modulation is basically similar to PWM. Difference is that duty cycle in this case is always constant (D = const.), but frequency of switching signal is varying ($f_{sw}$ = var.). Based on the change of control voltage $u_{reg}$, the switching frequency of PWM signal is being changed. In the case of analog circuits the frequency modulation is realized through utilization of voltage control oscillator (VCO). On the other hand, in the case of digital implementation the VCO is not required, because frequency of timer is able to be changed directly.

These previously mentioned ways of regulation of electric quantities in power converters are basic types, whereby also their combination is being utilized in practical applications.

## 3. LLC converter - properties and principle of operation

LLC resonant converter is multi-resonant converter and is characterized by its unique DC - gain characteristic, which has two resonant frequencies ($f_{r1}$ and $f_{r2}$). This converter has several advantages compared to standard serial LC resonant topology. One of them is possibility of stable regulation of output voltage in a wide range of input voltages together with the change of output power from 1% to 100%. The next advantage is achievement of ZVS switching mode during various operational modes. LLC resonant converter is composed of three functional parts (Fig. 1a). It deals about pulse generator, resonant circuit with high-frequency transformer and rectifier with capacitive filter. Operation of LLC converter in different operational modes is described by DC gain characteristic (Fig. 1b),

which should be divided into ZVS and/or ZCS region. ZVS region in dependency on the switching frequency can be further divided into:
- region with switching frequency equal to resonant ($f_{sw} = f_{r1}$)
- region1 with switching frequency higher than resonant ($f_{sw} > f_{r1}$)
- region2 with switching frequency lower than resonant ($f_{sw} < f_{r1}$)
- region3 with switching frequency lower than resonant and with voltage gain < 1

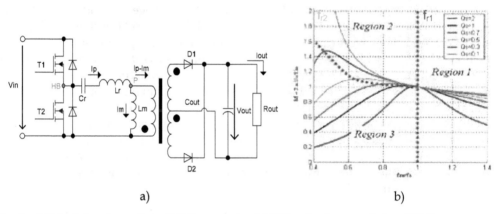

a)                                                              b)

Fig. 1. a) Principle schematics of LLC converter, b) DC-gain characteristic of LLC converter

According to the operational modes of resonant converters the operation of LLC resonant converter is rather difficult. The principal waveforms of transformer and output diode during each operating mode are shown in Fig. 2. The impedance of series resonant circuit at the resonant frequency is equal to zero. Therefore the reflected output voltage is equal to the input voltage, what is described by the unity of voltage gain thus the circuit then operates optimally. LLC resonant converter can achieve greater gain, lower or equal to 1. If the switching frequency is less than the resonant frequency, magnetization inductance is involved into the resonance of the circuit so the converter can deliver higher gain.

For analysis of operation of LLC converter it is necessary to exactly examine voltage gain characteristic. Voltage gain characteristic is possible to be made from equivalent circuit of converter. Output part together with rectifier and load is replaced by equivalent output resistance which is marked as $R_{AC}$.

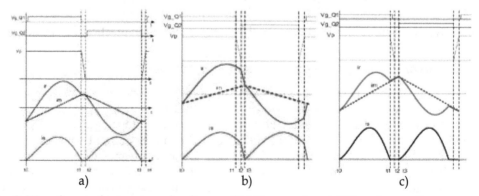

a)                                  b)                                  c)

Fig. 2. Waveforms of currents and voltages of LLC converter at different operating conditions a) $f_{sw} = f_{r1}$, b) $f_{sw} > f_{r1}$, c) $f_{sw} < f_{r1}$

Considering, that converter is working with almost resonant frequency, only first harmonic is transferred through resonant circuit, and therefore it is possible to utilize simple approximation, where rectangular pulses are substituted by sinusoidal waveforms.

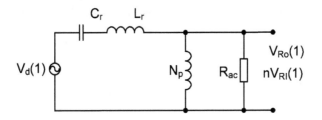

Fig. 3. Equivalent circuit of LLC converter

For voltage expressions of this equivalent circuit next formulas are valid:

$$V_{RI(1)} = \frac{4 \cdot V_o}{\pi} \cdot \sin(\omega t) \tag{2}$$

$$V_{d(1)} = \frac{4}{\pi} \cdot \frac{V_{in}}{2} \cdot \sin(\omega t) \tag{3}$$

where $V_{RI}$ is voltage at the output of resonant circuit, and $V_d$ is first harmonic part of supply voltage.

The equivalent circuit of converter is then similar to that which is shown on Fig.3. Equation of voltage gain characteristic is then equal to:

$$M = \frac{2nV_o}{V_{in}} = \left| \frac{\left( \frac{\omega^2}{\omega_{r1}^2} \right) \cdot \sqrt{m \cdot (m-1)}}{\left( \frac{\omega^2}{\omega_{r2}^2} - 1 \right) + j \cdot \left( \frac{\omega}{\omega_{r1}} \right) \cdot \left( \frac{\omega^2}{\omega_{r1}^2} - 1 \right) \cdot (m-1) \cdot Q} \right| \tag{4}$$

where:

$$f_{r1} = \frac{1}{2\pi \cdot \sqrt{L_R \cdot C_R}} \tag{5}$$

$$f_{r2} = \frac{1}{2\pi \sqrt{(L_M + L_R) \cdot C_R}} \tag{6}$$

$$Q = \frac{\sqrt{\frac{L_R}{C_R}}}{R_{AC}} \tag{7}$$

$$R_{AC} = \frac{8 \cdot n^2}{\pi^2} \cdot R_O \tag{8}$$

Voltage gain characteristic of proposed converter is shown on Fig. 4. X-axis is defined as switching frequency, or better said frequency of input sinusoidal voltage. Ratio between output voltage and input voltage is marked on y-axis. Each wave is related to different loads of converter (quality factors Q). Main parameters of converter are: $V_{in} = 400$ V, $V_{out} = 60$ V, $L_r = 6.6$ µH, $C_r = 900$ nF, $L_m = 36$ µH, $L_{r2} = 3$ µH, $R_{omin} = 2.4$ Ω,   $I_{max} = 25$ A

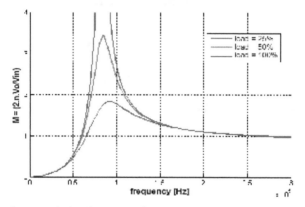

Fig. 4. Voltage gain characteristic of proposed converter.

### 3.1 Regulation of LLC converter

The regulation algorithm for the LLC converter maintains the requested value of output voltage, during the change of input voltage or output load. The LLC converter is controlled by variations of switching frequency, so the active part of the regulator is a voltage controlled oscillator (VCO) in analog or digital (software) form. The change of the output frequency is based on the input voltage of the VCO. As mentioned above, the switching frequencies of the LLC converter are between $f_{r1}$ and $f_{r2}$ during normal operation. The regulation algorithm works as follows: The normal operating point (Fig. 5a point 1) is characterized by $V_{IN}$=400 V, $U_{OUT}$=60 V. If the input voltage drops below 400 V, the regulation algorithm reacts in form of switching frequency reduction, so the operating point of the converter moves from the point 1 to the point 2. In this point, the converter works with the increased gain to maintain the output voltage at the constant value (60 V). The minimal input voltage is defined as 325V, so it is not possible to maintain the output voltage on 60V below this value.

If the input voltage rises back to 400V, the control system increases the switching frequency - voltage gain decreases and the operating point of the converter moves back to the point 1. The next operating condition for the LLC converter is variation of the output load. Fig. 5b shows the point 1, where the converter works with full load, the operating point is on the curve with Q=100%. When the output load drops, the operating point moves from the curve with Q=100% to the curve with Q=10%, which means that for the constant switching frequency, the output voltage rises to 60% above the nominal value, so the regulation algorithm rises the switching frequency which leads to lower voltage gain and output voltage decrease to nominal value.

Specific case of operation is the start of the converter, when the switching frequency is higher than the resonant frequency $f_{r1}$ to avoid current inrush during the initial charge of the output capacitor $C_{OUT}$. Generally, for the soft start, the switching frequency is two or three times higher than the resonant frequency. Voltage gain on this frequency is lower (Fig. 4), so the charging current of the output capacitor stays on an acceptable level.

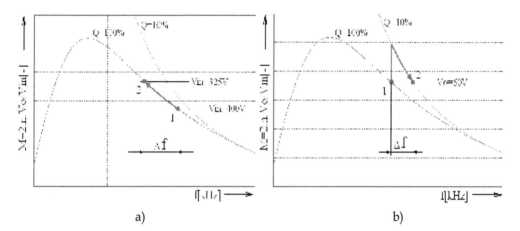

a)                                                                      b)

Fig. 5. Regulation of output voltage a) during input supply voltage decrease, b) during load variations

## 3.2 Simulation of LLC converter in OrCad PSPICE

For verification and mathematical model design, first, the simulation in Orcad PSPICE has been made. Data acquired from simulations in PSPICE are used in MATLAB for design of mathematical model. For best performance and accuracy of switching waveforms, all parasitic components were included into simulation.

For power transistors the IPW60R165CP model was used, and MUR10015CT was used for output diodes. For transformer, the leakage inductance with all parasitic components was also simulated.

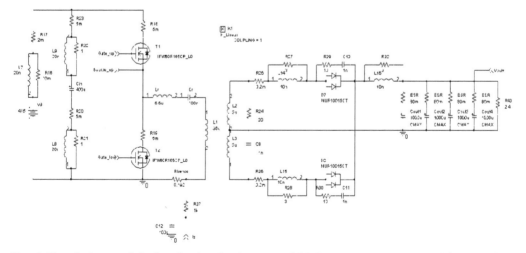

Fig. 6. Simulation model of main circuit of proposed LLC resonant converter with parasitic components

In Fig.7 the basic waveforms of LLC converter are shown. This simulation experiment was performed with full input voltage 425V and with output power 1kW. Fig. 8 shows another waveforms, with lower input voltage at 325V and output power of 1kW. All operation conditions including soft start and overload were also verified in simulation experiments.

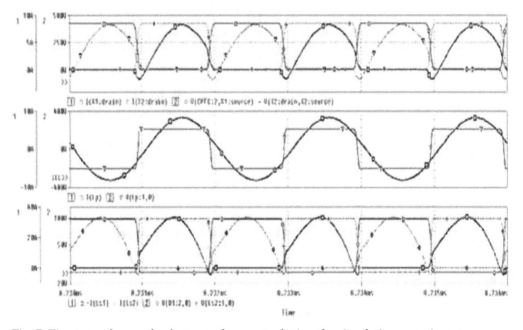

Fig. 7. Time waveforms of voltages and currents during the simulation experiment:
Uin = 425V, Pout = 1008W (from top:  transistor T1and T2, transformer primary side, output
diode D1 and D2)

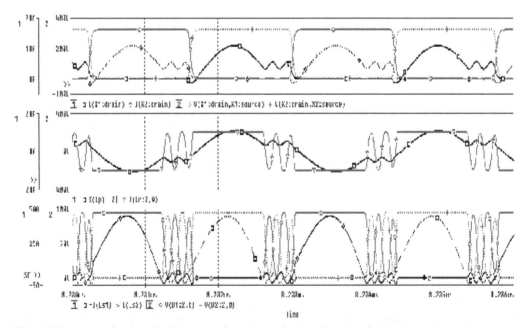

Fig. 8. Time waveforms of voltages and currents during the simulation experiment:
Uin = 325V, Pout = 1008W (from top - transistor X1a X2, transformer primary side, output
diode D1 and D2)

From simulations above, it's clear to say, that design procedure was very accurate and results are showing acceptable performance.

## 4. Transfer function for controller

In previous sections, a operating principle of LLC converter was briefly mentioned, together with simulation of main circuit. Accurate simulation is critical for design and optimization of the control system for the LLC converter. For the design of control structure and the actual control system, a transfer function is necessary, which will describe response of output voltage and current to the change of the control variable.

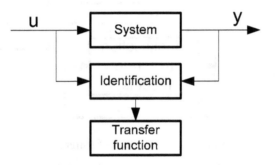

Fig. 9. Process of transfer function acquiring

Mathematical model of LLC converter requires control-to-output transfer function. Based on this transfer function, digital controller can be obtained. For PWM converters, standard "averaged" methods can be used for revealing the transfer function of converter. One of method with good results is "direct circuit averaging". This method can be easily implemented for standard PWM converters such as boost, buck, flyback etc. Transfer function obtained with this method has duty cycle as an input value and output voltage as an output value.

Unlike the PWM converters, the control transfer function of frequency controlled resonant converters cannot be obtained by state space averaging method, due to different ways of energy processing. While state space averaging methods eliminates the information about switching frequency, they cannot predict dynamic properties of resonant converters, so the proper control-to-output transfer function cannot be evaluated.

There is a several methods for solving this problems, but some of them are too simplified and idealized, others are too complex and difficult to use. In this paper, new simulation based method for revealing the control transfer function is proposed. This method is based on PSPICE simulation, which was discussed in previous chapter, and use of System Identification Toolbox in MATLAB environment. First, the simulation of main circuit in PSPICE must be created. Using of PSPICE simulation, the dependency of output (voltage, current) on input (switching frequency, duty cycle) can be simulated.

Another option for creation of the mathematical model is use of block called SLPS in MATLAB. This block creates interface between circuit model in PSPICE and mathematical models in MATLAB environment. Data acquired from simulation in PSPICE are used in MATLAB for creation of transfer function of LLC multiresonant converter. For obtaining of the transfer function in s-domain, the MATLAB System Identification Toolbox was used.

Whole process is shown on Fig. 10. First, the simulation of dynamic step response of LLC converter on control value (switching frequency) was created in OrCad PSPICE environment. Result from this simulation was dynamic output voltage response (output value) on step of switching frequency (input value). This simulated results from OrCad PSPICE, were used in MATLAB as input values for identification of this dynamic system.

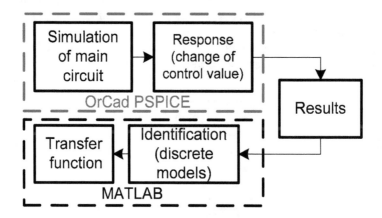

Fig. 10. Simulation based method for identification of transfer function

Use of System Identification Toolbox offers several models for identification of dynamic system. With use of different models from System Identification Toolbox (SIT), identification of all converters is possible.

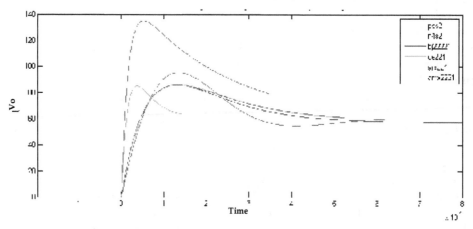

Fig. 11. Accuracy of some model types from SIT with different degree (step response)

System Identification Toolbox includes number of models in continuous or discrete form, which can be used for identification of systems: ARX (Auto Regresive Exogeneous Input Model), ARMAX (Auto Regresive Moving Average Exogeneous Input Model) OE (Output Error Model), BJ (Box-Jenkins Model), SS (State-Space Model). Accuracy of the models depends on degree of polynomials used in transfer function. Fig. 11 shows accuracy of identified transfer functions for different model types from simulated results of LLC converter. All models are in discrete form (z-domain), so the exact specification of sampling interval is necessary, which means, that the integration step which was used in OrCad PSPICE, must be also used for MATLAB. From Fig. 11 can be seen, that the best accuracy has ARX and ARMAX model.

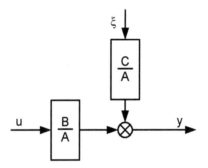

Fig. 12. Structure of ARMAX model

The ARMAX model, whose structure is on Fig. 12, is similar to ARX model, but with better accuracy. It extends the ARX model with polynomial C(z) - equation of average value of perturbation function. The ARMAX model is suitable for those systems, where perturbation function affects input variable u(n). Searched transfer function has the same form as in ARX model, but the computing algorithm is different. The equation for ARMAX model is as follows:

$$A(z)y(n) = \left(B(z)u(n-k) + C(z)\xi(n)\right) \tag{9}$$

Structure of the model is shown on Fig. 12. After adjusting, the form of Eq. (10) is:

$$y(n) = \frac{1}{A(z)}\left(B(z)u(n-nk) + C(z)\xi(n)\right) \tag{10}$$

Due to best accuracy of ARMAX model, only this model was used for revealing the transfer function. Another parameter necessary for the transfer function revealing, is degree of its polynomials. Fig. 13 shows accuracy of ARMAX model with different degree. Best accuracy offers model of 2nd degree, which means, that polynomials A(z), and C(z) are just 2nd degree. We can say, that the control to output transfer function of LLC converter has same degree as ARMAX model.

### 4.1 Transfer function
As mentioned above, the sampling interval used for models in SIT must be equivalent with integration step used for simulation in OrCad PSPICE. If sampling interval used in SIT differs from integration step in PSPICE, the dynamic results (from step response) will have different time representation in MATLAB, so the searched transfer function will have incorrect form.

Requested discrete transfer function is in form:

$$G(z) = \frac{B(z)}{A(z)} = \frac{b_0 + b_1 z^{-1} + b_2 z^{-2} + \ldots + b_{nb} z^{-nb}}{a_0 + a_1 z^{-1} + a_2 z^{-2} + \ldots + a_{na} z^{-na}} \tag{11}$$

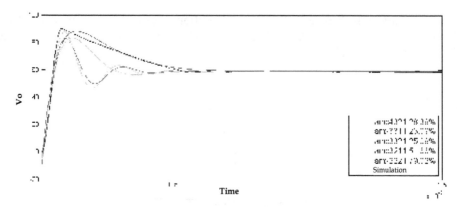

Fig. 13. Accuracy of ARMAX models with different degree (up to down 4,3,2)

For better view, Fig.14 shows comparison of ARMAX 2221 model, used for transfer function, with simulated dynamic step response from OrCad PSPICE.

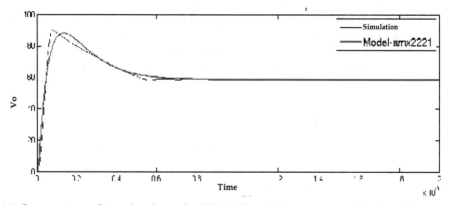

Fig. 14. Comparison of simulated results (PSPICE) with ARMAX model of 2nd degree (MATLAB)

For LLC converter, whose waveforms are in Fig. 7 and Fig. 8, resulting transfer function is in Eq. (12).

$$G(s) = \frac{2.054 \cdot 10^6 s + 6.895 \cdot 10^9}{s^2 + 2.29 \cdot 10^4 s + 1.176 \cdot 10^8} \tag{12}$$

After acquiring of transfer function, we can obtain basic mathematical model of LLC converter. This mathematical model can be used for design of discrete model, with all effects from discrete control system. The following figures show the bode characteristics of the systems marked $G_{dlyz3}$ and $G_{dlyz4}$ with discrete controllers. The design of controllers was made in Matlab using the SISOTOOL. This tool allows to design not only regulators, but also analyze the characteristics and quality of the proposed control circuit.

## 5. Design of discrete controller

Due to implementation in microprocessor, the controller must be in discrete form. There are two ways to design a discrete controller - design by emulation and direct digital design.

In the design by emulation approach, also known as digital redesign method, first an analog controller is designed in the continuous domain, by ignoring the effects of sampling and hold of A/D converter and computing delay of microprocessor. In next step, the controller can be converted into discrete-time domain by one of discretization method. This approach is good for systems of lower degree, but in discrete systems of higher degree, the transient responses does not reflect the required values because of ignoring sample and hold and computation delay effects.

On the other side, direct digital design approach offers design of controller directly in z-domain, without conversion, including effects of A/D converter and microprocessor. Block diagram of this approach is on Fig. 15.

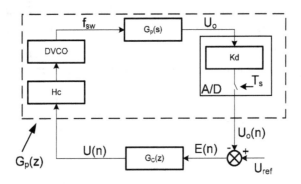

Fig. 15. Direct digital design control loop

Sampling of measured value with A/D converter can be represented by ideal Zero-Order-Hold block with sample time Ts. Gain of A/D converter is represented by block $K_{adc}$. Computing delay of microprocessor, also with delay from PWM module are represented by block $T_{comp}$. A/D converter with PWM module together form a sampling-and-hold device. Sample and hold block brings additional time delay of Ts/2 and phase lag of ωTs/2, which means, that reconstructed signal has time or phase lag. Block $T_{comp}$ represents delay between conversion of A/D converter and PWM duty cycle or modulo update. Time between this two events is necessary for computing the values for PWM block. Discrete transfer function of whole converter including Zero-Order block, Sample-and-Hold effect and gain of A/D converter is:

$$G\ (z) = Z\left\{\frac{1}{s}\left(1 - e^{sT_s}\right) \cdot H_C(s) \cdot G_P(s) \cdot K_d\right\}$$

Fig. 16 shows effect of sampling interval with computing delay on stability of closed control loop. Tab. 1 shows transfer functions with different sampling intervals and different computational delays. Sampling times were used from A/D converter included in DSC 56F8013, the computing times were used from same processor. Application was for digital control system for 200kHz LLC multiresonant converter. In this system discrete regulator of third order was used.

For design of controller in z-domain, all above mentioned delays must be taken into account. With use of MATLAB Siso Design Tool, the proper discrete controller can be designed. Advantage of this tool is possibility of direct placing of zeroes and poles of controller on bode diagram of closed or open loop. After placing the poles or zeros of controller, the different responses of closed control loop can be displayed for verification.

| Sampling time $T_s$ | Computing delay $T_{comp}$ | Mark |
|---|---|---|
| 5µs | 0 | Gz1 |
| 10µs | 0 | Gz2 |
| 5µs | 3µs | Gdlyz1 |
| 10µs | 3µs | Gdlyz2 |
| 5µs | 6µs | Gdlyz3 |
| 10µs | 6µs | Gdlyz4 |

Table 1. Sampling times and computational delays for transfer functions in Fig. 16

From Fig. 16 is clear, that computing delay has significant effect on stability of the control loop. With rising computing and sampling time stability of system drops.

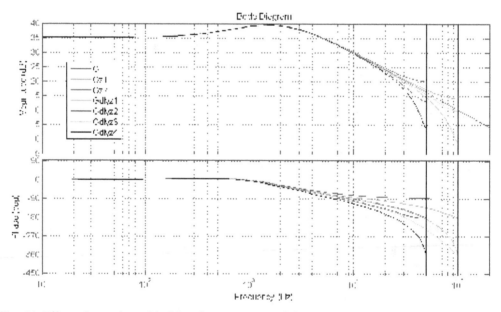

Fig. 16. Effect of sample and hold and computation delay on stability of control loop

Fig. 17 shows the system $G_{dlyz3}$ with poles and zeros of the regulated system (blue) and regulator (red). Fig. 18 shows the closed-loop response of the system to a unitary jump. Time of control settling is about 70 microseconds, overshoot as high as 17%.
Transfer function of designed controller is in form:

$$G_c(z) = 0.041204 \cdot \frac{z - 0.886}{z - 0.995} \tag{13}$$

Fig.19 shows the bode characteristics of system marked $G_{dlyz4}$, with the proposed controller. Poles and zeros of the transfer function of the regulated system are shown blue, poles and zeros of the controller are shown red. As in the previous case, in Fig. 20 a closed control loop response to unitary jump is shown. Time of control settling is 100 microseconds, overshoot is 20%, which is caused by a longer interval sampling interval of system Gdlyz4.

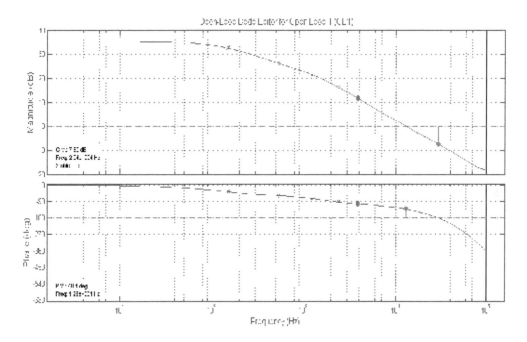

Fig. 17. Design of controller using transfer function of open loop for system $G_{dlyz3}$

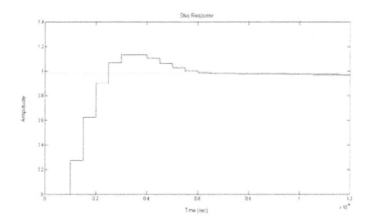

Fig. 18. Closed-loop response with designed controller for unitary jump

Both designed controllers controls at 97% of requested value, due to the reduction of regulators gain, because of limitation of overshoot below 20% of the requested value. The elimination of this problem was realized by increasing the requested value of the output voltage by 3%

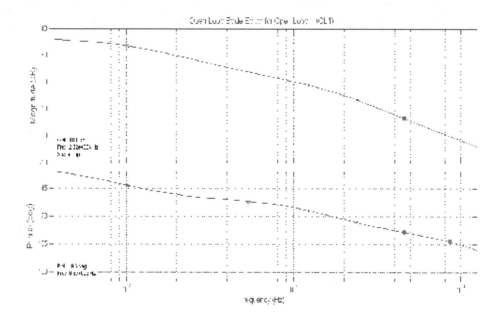

Fig. 19. Design of controller using transfer function of open loop

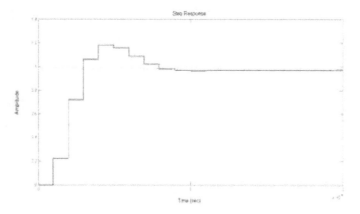

Fig. 20. Closed-loop response with designed controller for unitary jump

Transfer function of designed controller is in form:

$$G_C = 0.028539 \cdot \frac{z - 0.748}{z - 0.994} \tag{14}$$

SISOTOOL enables also the analysis of the quality and stability of the regulatory circuit. The advantage is, that information on the quality and stability control is available directly during the design procedure of the controller. As seen from figures 4.7 and 4.9 the systems with closed loop are stable , with safety margin about 50°.

### 5.1 Implementation in DSC

Discrete controller proposed in previous chapter was implemented into 16b digital signal controller (DSC) Freescale 56F8013, which is primary designed for motor and converter control. Advantage of this microprocessor are high performance peripherals which can operate with 96Mhz clock frequency. On the other side, disadvantage of this processor is low core frequency - 32 MHz, and fixed point arithmetic. For better performance, fraction arithmetic with intrinsic functions were used in this DSC.

Block scheme of digital control system with DSC 56F8013 on 200 kHz LLC converter is on Fig. 21. For better performance and lower amount of additional hardware, the concept with microprocessor on secondary side of converter was used. Also, new available fast digital isolators were used instead of standard optocouplers.

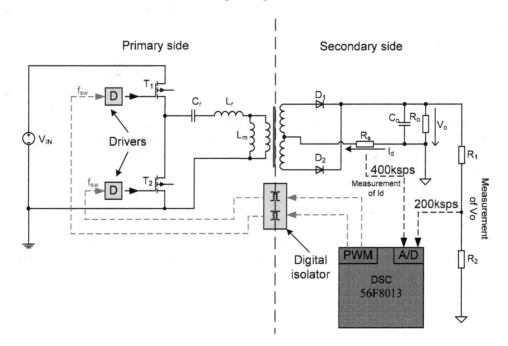

Fig. 21. Block scheme of full digital control system with DSC 56F8013

| Processor | Bits | Core frequency | A/D sampling time | Computing of control loop | Computing of current value |
|-----------|------|----------------|-------------------|---------------------------|----------------------------|
| 56F8013 | 16b | 32Mhz | 1.126µs | 2.98µs | 3.01µs |
| ColdFireV1 | 32b | 50,3Mhz | 2.252µs | 1.8µs | 1.12µs |

Table 2. Sampling times and computational delays for two processors

Another option for implementation is use of 32b microprocessor ColdFire V1 which offers better computing performance, but sampling time of A/D converter is twice as in DSC

56F8013. For measurement of output voltage, this time is sufficient, but for current sensing, the A/D converter on ColdFire is too slow. This problem was eliminated by use of special algorithm for computing of diode current from value of output voltage, value of load and ripple of output voltage. This algorithm is based on computing of output current from output voltage drop, during load connection. Detailed method is described in. Times required for computing on both processors are in Tab. 2.

## 6. Conclusion

The method for creation of mathematical model discussed in this chapter is based on circuit model of LLC converter in OrCAD PSPICE and use of MATLAB System Identification Toolbox. This approach eliminates the problems of transfer function obtaining, whose results from use of standard "averaging" methods. Instead of that, this approach uses simulation in OrCad PSPICE, based on circuit model of converter, for acquiring the results and subsequent use of MATLAB for creation of accurate control to output transfer function (in s- or z- domain). Advantage of this method is in use of all parasitic components (resistors, capacitances, inductors) in OrCad PSPICE simulation, so the transfer function is very accurate. Also, the approach for identification of the system, proposed in this paper, uses new, discrete model based method. This method allows rapid acquiring of mathematical model of converter with use of computing power of modern PC. Another advantage of this method is possibility of use for any type of converter without limitation in structure of converter.

## 7. Acknowledgement

Results of this work were made with support of national grant agency APVV project No. 0535-07 and to R&D operational program Centre of excellence of power electronics systems and materials for their components No. OPVaV-2008/01-SORO, ITMS 2622012003 funded by European regional development fund (ERDF).

## 8. References

Maksimovic, D. Zane, R. Erickson, R. (2004). Impact of Digital Control in Power Electronic, *Proceedings of 2004 International Symposium on Power Semiconductor Devices*, ISBN 4-88686-060-5, Kitakyushu, May 2004

Moudgalya, K. N. (2007). In: *Digital Control*, John Wiley and Sons, pp.157-237, John Wiley and Sons, ISBN 978-0-470-03144-5, Chichester

Hangseok, Ch. (2007). Analysis and design of LLC resonant converter with integrated transformer, *Proceedings of Applied Power Electronics Conference*, ISBN 1-4244-0714-1/07, Anaheim USA, March 2007

Hargas, L. Hrianka, M. Lakatos, J. Koniar, D. (2010). Heat fields modelling and verification of electronic parts of mechatronics systems, *Metalurgija (Metallurgy)*, vol. 49, (February 2010), ISSN 1334-2576

Frivaldský, M. Drgona, P. Prikopova, A. (2009). Design and modeling of 200kHz 1,5kW LLC power semiconductor resonant converter, *Proceedings of Applied Electronics*, ISBN 978-80-7043-781-0, September 2009

Frivaldský, M. Drgoňa, P. Špánik, P. (2009). Optimization of transistor´s hard switched commutation mode in high – power, high frequency application, *Proceedings of 15th International Conference on ELECTRICAL DRIVES and POWER ELECTRONICS*, ISBN 953-6037-55-1, Dubrovnik Croatia, October 2009

**6**

# Linear Variable Differential Transformer Design and Verification using MATLAB and Finite Element Analysis

Lutfi Al-Sharif, Mohammad Kilani, Sinan Taifour, Abdullah Jamal Issa, Eyas Al-Qaisi, Fadi Awni Eleiwi and Omar Nabil Kamal
*Mechatronics Engineering Department, University of Jordan*
*Jordan*

## 1. Introduction

The linear variable differential transformer is one of the most widely used transducers for measuring linear displacement. It offers many advantages over potentio-metric linear transducers such as frictionless measurement, infinite mechanical life, excellent resolution and good repeatability (Herceg, 1972). Its main disadvantages are its dynamic response and the effects of the exciting frequency. General guidelines regarding the selection of an LVDT for a certain application can be found in (Herceg, 2006).

The LVDT is also used as a secondary transducer in various measurement systems. A primary transducer is used to convert the measurand into a displacement. The LVDT is then used to measure that displacement. Examples are:

1. Pressure measurement whereby the displacement of a diaphragm or Bourdon tube is detected by the LVDT (e.g., diaphragm type pressure transducer, (Daly *et al.*, 1984)).
2. Acceleration measurement whereby the displacement of a mass is measured by the LVDT (e.g., LVDT used within an accelerometer, (Morris, 2001).
3. Force measurement whereby the displacement of an elastic element subjected to the force is measured by the LVDT (e.g., ring type load cell, (Daly *et al.*, 1984)).

The classical method of LVDT analysis and design is based on the use of approximate equations as shown in (Herceg, 1972) and (Popovic *et al.*, 1999). These equations suffer from inaccuracy especially from end effects. More novel methods for design employ finite element methods (Syulski *et al.*, 1992), artificial neural networks (Mishra *et al.*, 2006) and (Mishra *et al.*, 2005). The dynamic response of the LVDT is discussed in (Doebelin, 2003). The LVDT has also been integrated into linear actuators (Wu *et al.*, 2008).

## 2. General overview

A diagram showing the dimensional parameters of the LVDT is shown in Figure 1 below. The important parameters that are taken into consideration in the design of the LVDT are listed below.

1. The length of the primary coil, $l_p$.
2. The length of the secondary coil, $l_s$. It is assumed that both secondaries have the same length.

3. The length of the core, $l_c$.
4. The diameter of the core $r_c$.
5. The inner diameter of the coils, $r_i$. It is assumed that the primary and both secondaries have the same inner radius.
6. The outer diameter of the coils, $r_o$. It is assumed that the primary and both secondaries have the same outer diameter.
7. The separation of the wires (centre to centre), $w_s$. It is important to note that the diameter of the wire itself is not taken into consideration and is assumed to be less than the separation (centre to centre). This is the best arrangement as it makes an allowance for the insulation of the wire. The diameter of the wire affects the effective overall resistance of the coil. As shown in Figure 1, it can be seen that the wires are assumed to be arranged in rows and columns.
8. The excitation frequency, $f$.
9. The primary current peak value, $I$. A sinusoidal waveform is applied to the primary in order to achieve a peak value of current $I$.
10. The separation between the coils, $c_s$. This parameter is important, as the former (spindle) on which the three coils are wound must have some thickness to retain its structural integrity.
11. The material of the rod has been assumed to be pure iron with a relative permeability $\mu_r$ of 14 000.

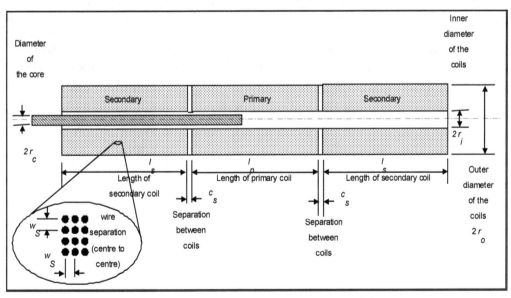

Fig. 1. Diagram showing the critical dimensions of the LVDT model

## 3. Objectives and methodology

The main objective of this piece of work is to develop a methodology for the design and verification of the linear variable differential transformer, using MATLAB to control and communicate with a magnetic finite element analysis tool. The methodology involves the following elements:

1. The capturing and coding of a number of rules of thumb that are used to find initial suitable values for the primary, secondaries and core length in relationship to the required stroke. These rules of thumb have been based on industrial experience.
2. The use of a finite element model (finite element magnetic tool) that is used to find the total flux linkage between the primary and the two secondaries based on a certain position of the core.
3. The use of MATLAB as a control tool to call the finite element modeling tool in order to find the output voltage of the two seconardies at each core position. MATLAB is then used to repeat this process until the full curve is produced.
4. A MATLAB based graphical user interface (GUI) has also been developed to act as a user friendly platform allowing the user to enter the required specification and to derive the output voltage characteristics.
5. Theoretical verification has been carried out, whereby the equation for the total flux linkage between two loops has been developed and then checked against the output of the finite element model to ensure that it is producing correct results.
6. In order to provide practical verification, an LVDT has then been built and the output measured and compared with the theoretical finite element outputs.

## 4. FEMM based model

The aim of the modeling methodology is to derive the transfer characteristic of an LVDT with certain dimensions and parameters. The transfer characteristic (or output characteristic) is a relationship between the displacement of the core and the output resultant dc voltage. It is assumed that the two ac signals from the two secondaries are processed by full wave rectifying them, smoothing the signals and then subtracting them.

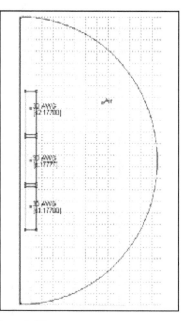

Fig. 2. Use of FEMM for LVDT modelling.

The FEMM (Finite Element Method, Magnetic) software is used to model the LVDT and find the total coupling flux from the primary to each of the secondaries at each position of the core. The total coupled flux is then divided by the primary core current to find the mutual inductance at this specific core position. By knowing the mutual inductance between the primary and each of the secondaries at a certain core position, the two output ac voltages can be found and hence the dc output voltage could be found. This process is then repeated for many other positions of the core, and the total transfer characteristic is then plotted.

The MATLAB software controls the whole process. It acts as the main controller to call the finite element magnetic tool and generate a point by point plot of the secondary output voltage at each point that corresponds to the position of the rod. Figure 2 shows the model of an LVDT within FEMM. It only shows half of the LVDT assuming symmetry between the two halves (left and right).

Using the FEMM software a large number of runs is carried out. These runs are automated using MATLAB. Starting values for all the parameters are used and these are referred to as the default parameters. Then one of the variable parameters is varied while all of the other parameters are kept constant.

| Symbol | Description | Default value | variable/fixed | Range |
|---|---|---|---|---|
| $l_p$ | length of the primary coil | 80 mm | variable | 26 to 133 mm (10 values) |
| $l_s$ | length of the secondary coil | 75 mm | variable | 36 to 125 mm (9 values) |
| $l_c$ | The length of the core | 130 mm | variable | 62 to 216 mm (9 values) |
| $r_c$ | The radius of the core | 4.5 mm | variable | 2.1 to 7.5 mm (9 values) |
| $r_i$ | The inner radius of the coils | 10 mm | variable | 4.8 to 16.7 mm (9 values) |
| $r_o$ | The outer radius of the coils | 35 mm | variable | 16.8 to 58.3 mm (9 values) |
| $w_s$ | The separation of the wires (centre to centre) | 0.3 mm | variable | 0.14 to 0.50 mm (9 values) |
| $f$ | The excitation frequency | 50 Hz | fixed | Not applicable |
| $I$ | The primary current peak value | 50 mA | fixed | Not applicable |
| $c_s$ | The separation between the coils | 5 mm | fixed | Not applicable |
| $\mu_r$ | Relative permeability of the core | 14 000 | fixed | Not applicable |

Table 1. Parameters of the LVDT modelling.

## 4.1 Generation of a set of curves

In order to generate a set of curves that show the change of the output characteristic with the change of certain design parameters, a number of curves have been generated as follows.

Each variable is changed 9 times within the range, while keeping all other parameters fixed. Figure 3, Figure 4, Figure 5, Figure 6, Figure 7, Figure 8 and Figure 9 show the effect on the transfer characteristic of changing the primary length, secondary length, core length, core radius, coil inner radius, coil outer radius and wire separation respectively.

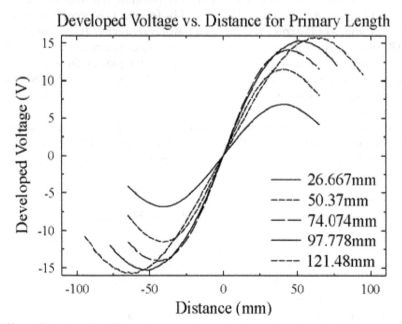

Fig. 3. Effect of the length of the primary coil on the transfer characteristic.

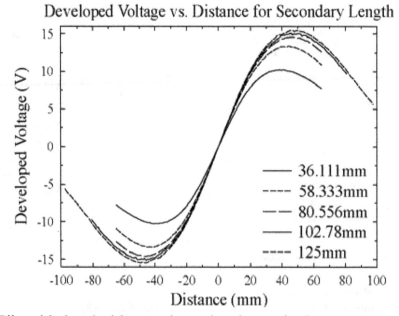

Fig. 4. Effect of the length of the secondary coil on the transfer characteristic.

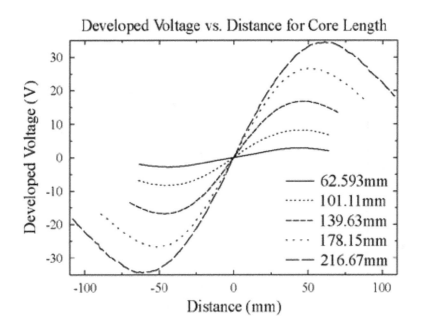

Fig. 5. Effect of the length of the core on the transfer characteristic.

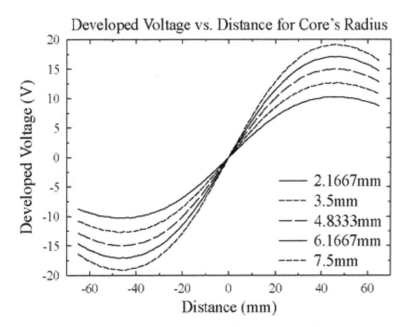

Fig. 6. Effect of the radius of the core on the transfer characteristic.

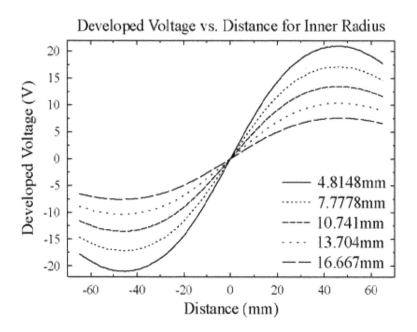

Fig. 7. Effect of the inner radius of the coils on the transfer characteristic.

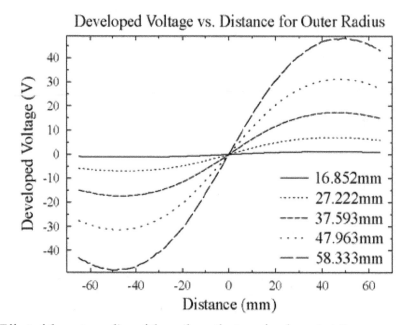

Fig. 8. Effect of the outer radius of the coils on the transfer characteristic.

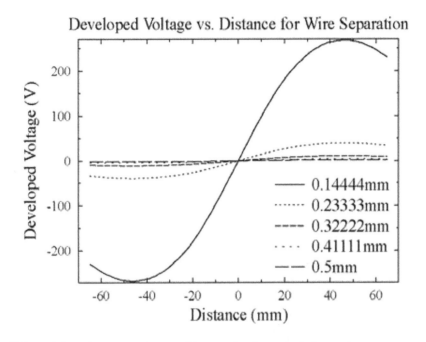

Fig. 9. Effect of the wire separation on the transfer characteristic.

MATLAB code has been used to automate the process of varying the parameters and to call the finite element modeling magnetic software. A parameter that is specific to the finite element software used is the so-called coarseness factor ($c_f$). Larger values of this variable result in finer graining. This is set to 18.

The two main characteristics of the LVDT that can be drawn from the graphs above are the sensitivity (mV/mm/V) and the stroke (mm). These are discussed in the next section.

### 4.2 Effect of parameters on stroke and sensitivity

The information gathered in the graphs generated in the last sub-section can be used to draw some general conclusions regarding the stroke and the sensitivity. The stroke is expressed in mm and the sensitivity is expressed in units of mV/mm/V (i.e., mV output volts dc for every mm core displacement for every voltage of excitation on the primary).

The effects of the primary length, secondary length, core length and outer coil radius on the stroke are shown in Figure 10, Figure 11, Figure 12 and Figure 13 respectively. It can be seen that the stroke increases with the length of the all of the four parameters, although it does flatten off in case of the secondary length.

The effects of the primary length, secondary length, core length and outer coil radius on the sensitivity have been shown in Figure 14, Figure 15, Figure 16 and Figure 17 respectively. The sensitivity is measured as the slope of the transfer characteristic at the null point.

It can seen that the sensitivity increases with the secondary length and the core length and decreases with the primary length and the outer radius of the coils.

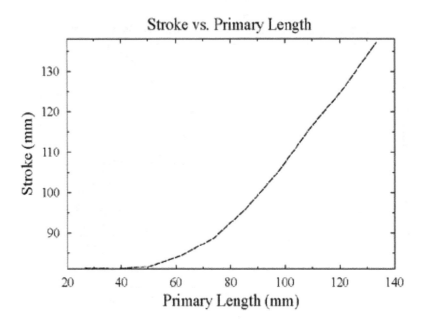

Fig. 10. Effect of the length of the primary coil on the stroke.

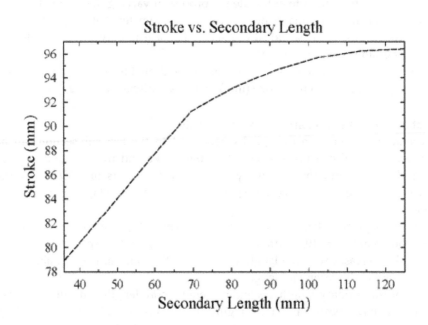

Fig. 11. Effect of the length of the secondary coil on the stroke.

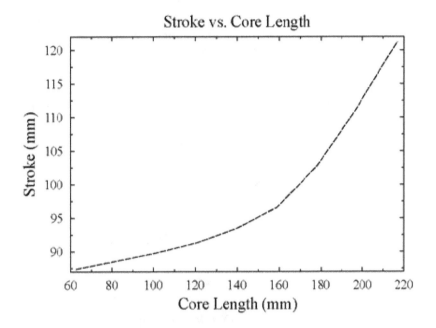

Fig. 12. Effect of the length of the core on the stroke.

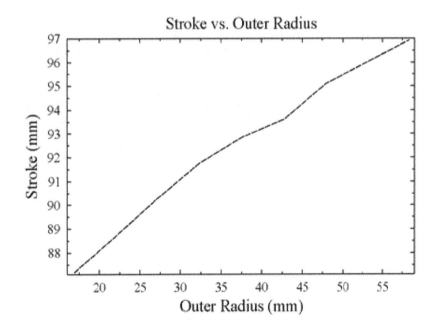

Fig. 13. Effect of the outer radius of the coils on the stroke.

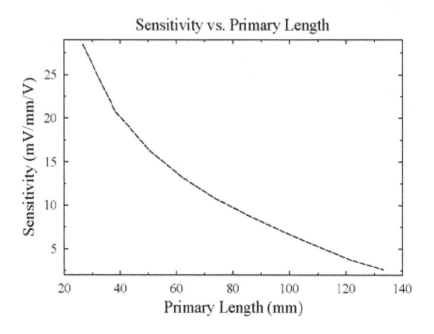

Fig. 14. Effect of the length of the primary coil on the sensitivity.

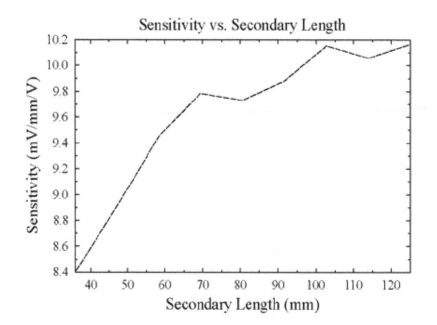

Fig. 15. Effect of the length of the secondary coil on the sensitivity.

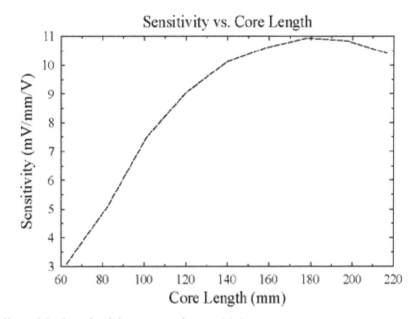

Fig. 16. Effect of the length of the core on the sensitivity.

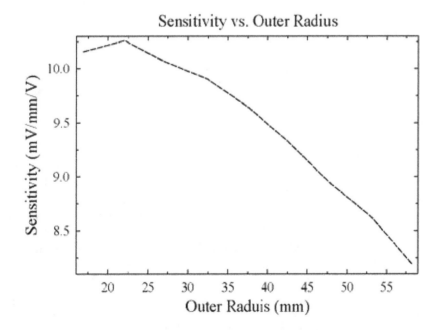

Fig. 17. Effect of the outer radius of the coils on the sensitivity.

## 5. Matlab GUI as a LVDT design tool

In order to automate the design process, a MATLAB graphical user interface (GUI) has been developed in order to provide a user friendly platform that allows the user to enter all the required parameters and run the LVDT. The software behind the GUI includes the rules of thumb, as well as suggested default values to the user. The output generated by the GUI is the output voltage curve plotted against the position of the rod. An example of a GUI screen is shown below.

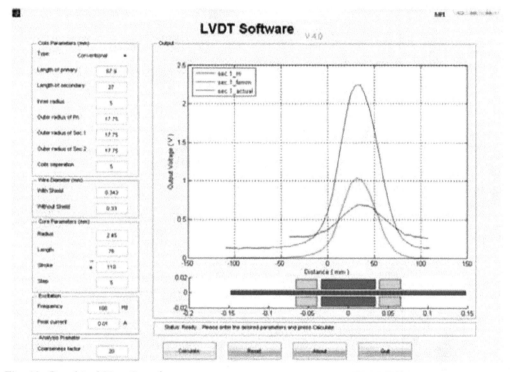

Fig. 18. Graphical User Interface.

## 6. Formulae for the flux linkage between two loops

In order to carry out a partial verification of the MATLAB/FEA model, an analytical method has been developed in order to find the expected output of each secondary based on the dimensions of the LVDT and the current in the primary. The basic model uses two concentric loops as shown Figure 19. One loop carries a certain current, and the flux linking the other loop is required. The aim is to find the total flux linked from one loop to another loop.

Let it be assumed that:
1.   The radius of the first loop is $A$
2.   The radius of the second loop is $B$
3.   The current in the first loop is i
4.   The two loops are concentric and separated by a distance $h$
5.   The permittivity of the medium separating them is $\mu$

The two loops lie in two parallel planes and are concentric as shown in Figure 19 below.

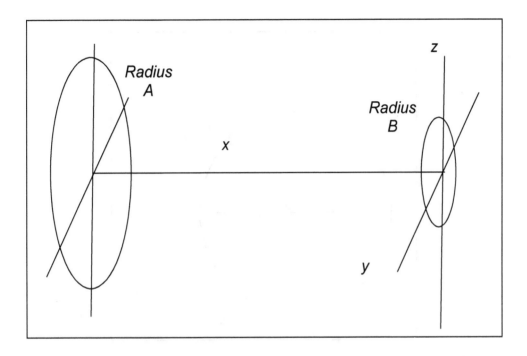

Fig. 19. General layout for the two loops.

Let us assume that loop A carries a current $i$. We will take a small length of current carrying conductor on loop A that has a length $dl$ (Figure 21).

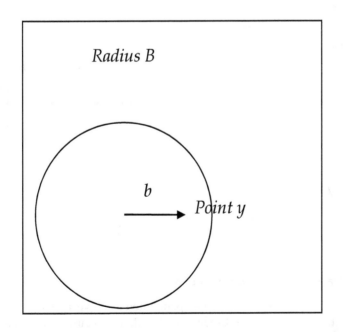

Fig. 20. Loop B

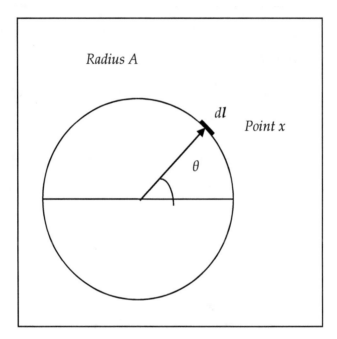

Fig. 21. Loop A.

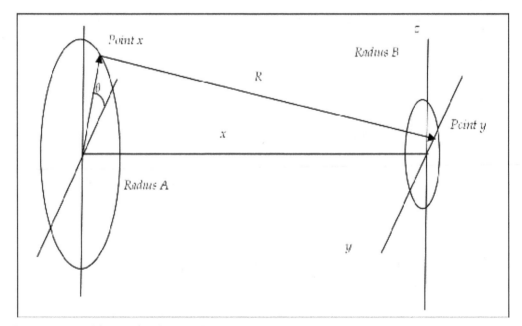

Fig. 22. General layout for the Biot-Savart Law.

We shall take a small section of the large loop, denoted as *dl* (vector) that carries a current *i*. The first step is to find the resultant magnetic flux density caused by a small section of the large loop dl (located at point *x*) at a point inside the smaller loop, denoted as point *y* (Figure 20). The point y has been taken inside the smaller loop on the y axis (without any loss of generality) at a distance b from the centre of the loop. The vector connecting points $x$ and $y$ represents the direction of the resultant magnetic flux density (Figure 22).

We shall denote the vector that connects points x to y as $R$. Using Biot-savart law gives:

$$\vec{dB} = \frac{\mu \cdot i \cdot \vec{dl} \times \vec{R}}{4 \cdot \pi \cdot R^3} \tag{1}$$

The coordinates of point $x$ are:

$$(0, A \cdot \cos\theta, A \cdot \sin\theta) \tag{2}$$

The coordinates of point $y$ are:

$$(h, b, 0) \tag{3}$$

The magnitude of R can be calculated as follows:

$$|R| = \sqrt{h^2 + (A \cdot \cos\theta - b)^2 + (A \cdot \sin\theta)^2} \tag{4}$$

$$|R| = \sqrt{h^2 + A^2 \cdot \cos^2\theta - 2bA \cdot \cos\theta + b^2 + A^2 \cdot \sin^2\theta}$$
$$= \sqrt{h^2 + A^2 + b^2 - 2bA \cdot \cos\theta} \tag{5}$$

This gives us the magnitude of R. We next find the components of the two vectors, $R$ and $dl$:

$$\vec{R} = \begin{bmatrix} h & b - A \cdot \cos\theta & -A \cdot \sin\theta \end{bmatrix} \tag{6}$$

$$\vec{dl} = \begin{bmatrix} 0 & -A \cdot \sin\theta \cdot d\theta & A \cdot \cos\theta \cdot d\theta \end{bmatrix} \tag{7}$$

We now turn to find the cross product of these two elements (note that we will only evaluate the x component as this is the component that is of interest to us).

$$\vec{dB_i} = \frac{\mu \cdot i \cdot d\theta \cdot \left(A^2 - A \cdot b \cdot \cos\theta\right)\vec{i}}{4 \cdot \pi \cdot \left(h^2 + A^2 + b^2 - 2 \cdot A \cdot b \cdot \cos\theta\right)^{\frac{3}{2}}} \tag{8}$$

We have taken the $x$ direction only as this is the direction that is perpendicular to the area of the smaller loop.

This effect is only caused by a small strip dl of the larger loop. In order to find the effect of the whole larger loop on the point $y$, we need to integrate around the larger loop. This is done as follows:

$$\vec{dB_{iloop}} = \int_0^{2\pi} \frac{\mu \cdot i \cdot \left(A^2 - A \cdot b \cdot \cos\theta\right)}{4 \cdot \pi \cdot \left(h^2 + A^2 + b^2 - 2 \cdot A \cdot b \cdot \cos\theta\right)^{\frac{3}{2}}} d\theta \, \vec{i} \tag{9}$$

If we now take an annulus of radius b inside the smaller loop, we can see that by symmetry, the value of the magnetic flux density component that is perpendicular to the area of the loop is constant everywhere on the annulus (Figure 23). We can now calculate the total flux that is passing through this annulus caused by the current $i$ in the large loop. We shall set the width of this annulus as $db$ and its radius as $b$.

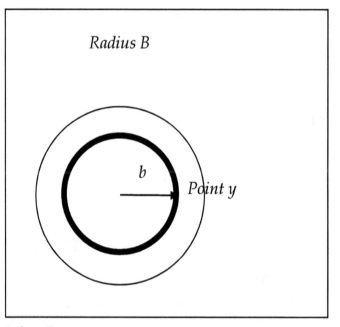

Fig. 23. Annulus in loop B.

So the total flux through the annulus can be found by multiplying the value of the magnetic flux density by the area of the annulus:

$$d\Phi_{annulus} = 2 \cdot \pi \cdot b \cdot \left( \int_0^{2\pi} \frac{\mu \cdot i \cdot \left( A^2 - A \cdot b \cdot \cos\theta \right)}{4 \cdot \pi \cdot \left( h^2 + A^2 + b^2 - 2 \cdot A \cdot b \cdot \cos\theta \right)^{\frac{3}{2}}} d\theta \right) db \tag{10}$$

So the total flux in the smaller loop can now be found by integrating over b from the value of b equal to 0 up to the radius of the smaller loop B.

$$\Phi = \int_0^B 2 \cdot \pi \cdot b \left( \int_0^{2\pi} \frac{\mu \cdot i \cdot \left( A^2 - A \cdot b \cdot \cos\theta \right)}{4 \cdot \pi \cdot \left( h^2 + A^2 + b^2 - 2 \cdot A \cdot b \cdot \cos\theta \right)^{\frac{3}{2}}} d\theta \right) db \tag{11}$$

$$\Phi = \int_0^B \left( \int_0^{2\pi} \frac{2 \cdot \pi \cdot \mu \cdot i \cdot b \cdot \left( A^2 - A \cdot b \cdot \cos\theta \right)}{4 \cdot \pi \cdot \left( h^2 + A^2 + b^2 - 2 \cdot A \cdot b \cdot \cos\theta \right)^{\frac{3}{2}}} d\theta \right) db \tag{12}$$

Simplifying gives the final result:

$$\Phi = \int_0^B \left( \int_0^{2\pi} \frac{\mu \cdot i \cdot b \cdot \left(A^2 - A \cdot b \cdot \cos\theta\right)}{2 \cdot \left(h^2 + A^2 + b^2 - 2 \cdot A \cdot b \cdot \cos\theta\right)^{\frac{3}{2}}} d\theta \right) db \tag{13}$$

The formula shown above cannot be solved analytically. MATLAB is used to evaluate the double integral.

Verification was carried out between the formula above implemented in MATLAB for a primary and a secondary without an inserted core and the output of FEMM. A numerical example is shown below. It shows excellent agreement between the analytical formula used in (13) and the output of the FEMM.

The parameters and results of the numerical example are as follows:

| | |
|---|---|
| Primary excitation current | 30 mA |
| Radius of primary loop | 25 mm |
| Radius of secondary loop | 15 mm |
| Distance between loops | 80 mm |
| Flux linkage | |
| FEMM | 1.37551e-11 Wb |
| MATLAB (double integration equation) | 1.3558e-11 Wb |

## 7. Practical verification

A practical verification of the results from the MATLAB/FEA model and the analytical equation within MATLAB was then carried out. A model of the LVDT is built and tested. The results are compared and verified with outputs from the models.

Figure 24 shows the output voltage as a function of the displacement of the core. This has been carried out at an excitation frequency of 100 Hz. As the frequency is increased it is noted that a larger discrepancy between the expected and actual outputs exists. The explanation for this discrepancy is the eddy current losses with the core (which is a conducting material as well as being ferromagnetic). Eddy current losses are not currently modeled and this will be necessary in order to fully quantify the expected eddy current losses.

## 8. Conclusions

A methodology has been developed that allows the user to design and verify the output of an LVDT. A finite element magnetic model in conjunction with MATLAB has been developed that allows the user to design an LVDT and produce the expected output characteristics. A graphical user interface has been added to the software to facilitate data entry and design visualization.

Analytical and practical verification has been carried out. Good agreement has been achieved via the analytical verification. However, a discrepancy has been noted in the practical verification caused by the Eddy current losses (as it seems to increase with the frequency). This requires further investigation and modeling.

Sensitivity analysis has been carried out on some of the design parameters. General conclusions have been drawn showing the effect of the primary coil length, secondary coil

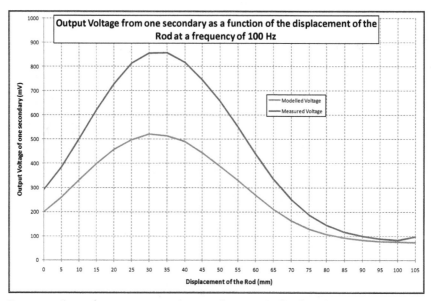

Fig. 24. Output voltage from one secondary as the core is displaced.

length, core length and outer radius of the coils on both the stroke and the sensitivity of the LVDT

## 9. Further work

Further work is still needed in the following areas:
1.  The seven variables that are varied are varied one at a time, keeping all the other parameters fixed. It is now the intention to vary all parameters at the same time to see the overall effect on the transfer characteristic.
2.  Four of the parameters are kept constant. Further work is needed to understand the effect of varying them on the transfer characteristic. Of particular importance is the effect of the frequency on the design (Herceg, 2006).
3.  The range of variation of the variables was restricted. More readings are needed outside the range used in order to draw more general conclusions.
4.  Variation of the variables is needed in pairs to see the relative importance of these variables. For example, the length of the core in relation to the primary as a ratio could be more important than the absolute values of the two variables.
5.  Further experimental verification is required. Moreover, modeling of the Eddy current losses is required in order to quantify the effect of the excitation frequency on the discrepancy between the expected output voltage and the measured output voltage.
6.  Further investigation is required into the possibility of combining a linear electromagnetic actuator with an LVDT in the same hardware as discussed in (Wu *et al.,* 1999). This offers a very compact closed loop linear electromagnetic actuator.

## 10. Acknowledgement

The practical verification and the analytical numerical example shown in this piece of work have been based on the graduation project work by a group of Mechatronics Enginering students at the University of Jordan. The members of the group are:

- "Mohammad Hussam ALDein" Omar Ali Bolad
- Anas Imad Ahmad Farouqa
- Anas Abdullah Ahmad Al-shoubaki
- Baha' Abd-Elrazzaq Ahmad Alsuradi
- Yazan Khalid Talib Al-diqs

## 11. References

Beckwith, Buck & Marangoni, (1982), Mechanical Measurements, Third Edition, 1982.

Daly, James; Riley, William; McConnell, Kennet, (1984), Instrumentation for Engineering Measurements, 2nd Edition, 1984, John Wiley & Sons.

Doebelin, E., (2003), Meaurement Systems: Application and Design, 5th Edition, McGraw Hill, 2003.

Herceg, Ed, (2006), Factors to Consider in Selecting and Specifying LVDT for applications, Nikkei Electronics Asia, March 2006.

Herceg, Edward, (1972), Handbook of Measurement and Control: An authoritative treatise on the theory and application of the LVDT, Schaevitz Engineering, 1972.

Mishra, SK & Panda, G, (2006), A novel method for designing LVDT and its comparison with conventional design, Proceedings of the 2006 IEEE Sensors Applications Symposium, pages: 129-134, 2006.

Mishra, SK; Panda, G; Das, DP; Pattanaik, SK & Meher, MR, (2005), A novel method of designing LVDT using artificial neural network, 2005 International Conference on Intelligent Sensing and Information Processing Proceedings, page 223-227, 2005.

Morris, Alan S., (2001), Measurement & Instrumentation Principles, Elsevier Butterworth Heinemann, 2001.

Popović, Dobrivoje; Vlacic, Ljubo, (1999), Mechatronics in Engineering Design and Product Development, CRC Press, 1999.

Syulski, J.K., Sykulska, E. and Hughes, S.T., (1992), Applications of Finite Element Modelling in LVDT Design, The International Journal for Computation and Mathematics in Electrical and Electronic Engineering, Vol. 11, No. 1, 73-76, James & James Science Publishers Ltd.

Wu, Shang-The; Mo, Szu-Chieh; Wu, Bo-Siou, (2008), An LVDT-based self-actuating displacement transducer, Science Direct, Sensors and Actuators A 141 (2008) 558-564.

# 7

# Application of Modern Optimal Control in Power System: Damping Detrimental Sub-Synchronous Oscillations

Iman Mohammad Hoseiny Naveh and Javad Sadeh
*Islamic Azad University, Gonabad Branch*
*Islamic Republic of Iran*

## 1. Introduction

The occurrence of several incidents in different countries during the seventies and the eighties promoted investigations into the cause of turbine-generator torsional excitation and the effect of the stimulated oscillations on the machine shaft. The best known incidents are the two shaft failures that occurred in the Mohave station in Nevada in 1970 and 1971, which were caused by sub-synchronous resonance (SSR) (Walker et al., 1975; Hall et al., 1976). A major concern associated with fixed series capacitor is the SSR phenomenon which arises as a result of the interaction between the compensated transmission line and turbine-generator shaft. This results in excessively high oscillatory torque on machine shaft causing their fatigue and damage. These failures were caused by sub-synchronous oscillations due to the SSR between the turbine-generator (T-G) shaft system and the series compensated transmission network. These incidents and others captured the attention of the industry at large and stimulated greater interest in the interaction between power plants and electric systems (IEEE committee report, 1992; IEEE Torsional Issues Working Group, 1997; Anderson et al., 1990; Begamudre, 1997).

Torsional interaction involves energy interchange between the turbine-generator and the electric network. Therefore, the analysis of SSR requires the representation of both the electromechanical dynamics of the generating unit and the electromagnetic dynamics of the transmission network. As a result, the dynamic system model used for SSR studies is of a higher order and greater stiffness than the models used for stability studies. Eigenvalue analysis is used in this research. Eigenvalue analysis is performed with the network and the generator modelled by a system of linear simultaneous differential equations. The differential and algebraic equations which describe the dynamic performance of the synchronous machine and the transmission network are, in general, nonlinear. For the purpose of stability analysis, these equations may be linearized by assuming that a disturbance is considered to be small. Small-signal analysis using linear techniques provides valuable information about the inherent dynamic characteristics of the power system and assists in its design (Cross et al., 1982; Parniani & Iravani, 1995).

In this research, two innovative methods are proposed to improve the performance of linear optimal control for mitigation of sub-synchronous resonance in power systems. At first, a technique is introduced based on shifting eigenvalues of the state matrix of system to the left hand-side of s plane. It is found that this proposed controller is an extended state of linear optimal controller with determined degree of stability. So this method is called extended optimal control. A proposed design, which is presented in this paper, has been developed in order to control of severe sub-synchronous oscillations in a nearby turbine-generator. The proposed strategy is tested on second benchmark model and compared with the optimal full-state feedback method by means of simulation. It is shown that this method creates more suitable damping for these oscillations.

In some of genuine applications, measurement of all state variables is impossible and uneconomic. Therefore in this chapter, another novel strategy is proposed by using optimal state feedback, based on the reduced – order observer structure. It was shown also that the Linear Observer Method can mitigate Sub-synchronous Oscillations (SSO) in power systems. The proposed methods are applied to the IEEE Second Benchmark system for SSR studies and the results are verified based on comparison with those obtained from digital computer simulation by MATLAB.

## 2. System model

The system under study is shown in Fig. 1. This is the IEEE Second benchmark model, with a fixed series capacitor connected to it. This system is adopted to explain and demonstrate applications of the proposed method for investigation of the single-machine torsional oscillations. The system includes a T-G unit which is connected through a radial series compensated line to an infinite bus. The rotating mechanical system of the T-G set is composed of two turbine sections, the generator rotor and a rotating exciter (Harb & Widyan, 2003).

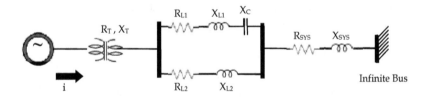

Fig. 1. Schematic diagram of the IEEE Second Benchmark System.

### 2.1 Electrical system
Using direct, quadrate (d-q axes) and Park's transformation, the complete mathematical model that describes the dynamics of the synchronous generator system:

$$\Delta \dot{X}_{Gen} = A_G \Delta X_{Gen} + B_{G1} \Delta U_1 + B_{G2} \Delta U_2 + B_{G3} \Delta U_3 + B_{G4} \Delta U_4 \tag{1}$$

$$\Delta y_{Gen} = C_G \Delta X_{Gen} \tag{2}$$

Where, $C_G$ is an identity matrix. The following state variables and input parameters are used in (1):

$$\Delta X_{Gen}^{T} = \begin{bmatrix} \Delta i_{fd} & \Delta i_d & \Delta i_{kd} & \Delta i_q & \Delta i_{kq} \end{bmatrix} \tag{3}$$

$$\Delta U_{Gen}{}^{T} = \begin{bmatrix} \Delta V_O & \Delta \delta_g & \Delta \omega_g & \Delta E \end{bmatrix}$$

(4)

Where, $\Delta V_O$ is variation of infinitive bus voltage. In addition to the synchronous generator, the system also contains the compensated transmission line. The linearized model of transmission line is given by:

$$\Delta \dot{X}_{Line} = A_{Line}\Delta X_{Line} + B_{Line}\Delta U_{Line}$$

(5)

$$\Delta X_{Line}{}^{T} = \begin{bmatrix} \Delta V_{Cd} & \Delta V_{Cq} \end{bmatrix}$$

(6)

$$\Delta U_{Line}{}^{T} = \begin{bmatrix} \Delta i_d & \Delta i_q \end{bmatrix}$$

(7)

To obtain the electrical system, we can combine (1–7). Finally we can illustrate electrical system by below equations:

$$\Delta \dot{X}_{El} = A_{El}\Delta X_{El} + B_{El}\Delta U_{El}$$

(8)

$$\Delta X_{El}{}^{T} = \begin{bmatrix} \Delta X_{Gen}{}^{T} & \Delta X_{Line}{}^{T} \end{bmatrix}$$

(9)

$$\Delta U_{El}{}^{T} = \begin{bmatrix} \Delta U_{Gen}{}^{T} \end{bmatrix}$$

(10)

## 2.2 Mechanical system

The shaft system of the T-G set is represented by four rigid masses. The linearized model of the shaft system, based on a mass-spring-damping model is:

$$\Delta \dot{X}_{Mech} = A_M \Delta X_{Mech} + B_{M1}\Delta U_{M1} + B_{M2}\Delta U_{M2}$$

(11)

$$\Delta X_{Mech}{}^{T} = \begin{bmatrix} \Delta \delta_1 & \Delta \omega_1 & \Delta \delta_g & \Delta \omega_g & \Delta \delta_2 & \Delta \omega_2 & \Delta \delta_3 & \Delta \omega_3 \end{bmatrix}$$

(12)

$$\Delta U_{Mech}{}^{T} = \begin{bmatrix} \Delta T_m & \Delta T_e \end{bmatrix}$$

(13)

The variation of electrical torque is denoted by $\Delta T_e$ and is given by:

$$\Delta T_e = P_1.\Delta i_{fd} + P_2.\Delta i_d + P_3.\Delta i_q + P_4.\Delta i_{kd} + P_5.\Delta i_{kq}$$

(14)

Parameters $P_1 - P_5$ can simplicity be founded by combination of electrical and mechanical system (Fig 2). Fig.3 illustrates the shaft system of the turbine-generator (T-G) in IEEE second benchmark model.

## 2.3 Combined power system model

The combined power system model is obtained by combining the linearized equations of the electrical system and mechanical system.

Let us define a state vector as $\Delta X_{Sys}{}^{T} = [\Delta X_{El}{}^{T} \ \Delta X_{Mech}{}^{T}]$. So we can write:

$$\Delta \dot{X}_{Sys} = A_{Sys}\Delta X_{Sys} + B_{Sys}\Delta U_{Sys} \tag{15}$$

$$\Delta U_{Sys}{}^{T} = \begin{bmatrix} \Delta T_m & \Delta E \end{bmatrix} \tag{16}$$

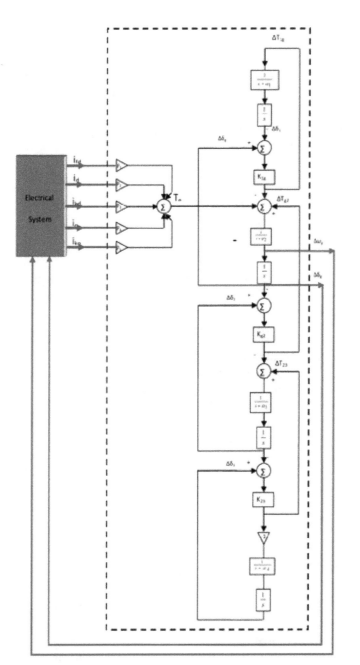

Fig. 2. Schematic diagram of calculation of $T_e$

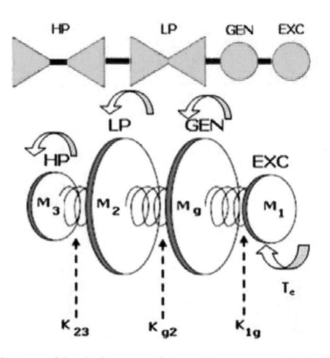

Fig. 3. Schematic diagram of the shaft system of the turbine-generator IEEE second benchmark model.

## 3. Modern optimal control

Optimal control must be employed in order to damp out the sub-synchronous oscillations resulting from the negatively damped mode. For the linear system, the control signal U which minimizes the performance index (Zhu et al., 1996; Patel & Munro, 1982; Khaki Sedigh, 2003; Ogata, 1990; Friedland, 1989; Kwakernaak, 1972):

$$J = \int [\Delta x_{Sys}{}^{T}(t) Q \Delta x_{Sys}(t) + \Delta u_{Sys}{}^{T} R_{\mu} \Delta u_{Sys}(t)]dt \qquad (17)$$

It is given by the feedback control law in terms of system states:

$$U(t) = -K \cdot \Delta x_{Sys}(t) \qquad (18)$$

$$K = R_{\mu}{}^{-1}.B_{Sys}{}^{T}.P \qquad (19)$$

Where P is the solution of Riccati equation:

$$A_{Sys}{}^{T}.P + P.A_{Sys} - P.B_{Sys}R_{\mu}{}^{-1}B_{Sys}{}^{T}.P + Q = 0 \qquad (20)$$

### 3.1 Extended optimal control
In this chapter, a strategy is proposed, based on shifting eigenvalues of the state matrix of system to the left hand-side of s plane, for damping all sub-synchronous torsional

oscillations. In order to have a complete research, optimal full state feedback control is designed and the results are compared with extended optimal control as proposed method. It is found that we can design linear optimal controller to obtain special degree of stability in optimal closed loop system by using this method. In the other hand, we can design a new controller that it transfers all of poles of optimal closed loop system to the left hand-side of – α (a real value) in S plane. For a linear system by (15), we can rewrite the performance index using proposed method:

$$J = \int e^{2\alpha t}.[\Delta x_{Sys}{}^{T}(t)Q\Delta x_{Sys}(t) + \Delta u_{Sys}{}^{T}R_{\mu}\Delta u_{Sys}(t)]dt \tag{21}$$

For the linear system the control signal U which minimizes the performance index is given by the feedback control law in terms of system states:

$$U_{\alpha}(t) = -K_{\alpha} \cdot \Delta x_{Sys}(t) \tag{22}$$

$$K_{\alpha} = R_{\mu}{}^{-1}.B_{Sys}{}^{T}.P_{\alpha} \tag{23}$$

Where Pα is the solution of Riccati equation in this case:

$$(A_{Sys} + \alpha I_{n})^{T}.P_{\alpha} + P_{\alpha}.(A_{Sys} + \alpha I_{n}) - P_{\alpha}.B_{Sys}R_{\mu}{}^{-1}B_{Sys}{}^{T}.P_{\alpha} + Q = 0 \tag{24}$$

Where $I_n$ is an n×n identity matrix. So the state matrix of optimal closed loop system (Aα) is obtained by below:

$$(A_{Sys} + \alpha I_{n})^{T}.P_{\alpha} + P_{\alpha}.(A_{Sys} + \alpha I_{n}) - P_{\alpha}.B_{Sys}R_{\mu}{}^{-1}B_{Sys}{}^{T}.P_{\alpha} + Q = 0 \tag{25}$$

In order to be asymptotically stable for Aα, we can write:

$$\lambda_{i}(A_{\alpha}) = \lambda_{i}(A_{Sys} + \alpha I_{n} - B_{Sys}.K_{\alpha}) = \lambda_{i}(A_{Sys} - B_{Sys}.K_{\alpha}) + \alpha \tag{26}$$

Where λi(Aα) is eigenvalues of Aα for i=1,2,...,n. Because of $Re[\lambda_{i}(A_{\alpha})] < 0$, we can write :

$$Re[\lambda_{i}(A_{\alpha})] = Re[\lambda_{i}(A_{Sys} - B_{Sys}.K_{\alpha}) + \alpha] = Re[\lambda_{i}(A_{Sys} - B_{Sys}.K_{\alpha})] + \alpha < 0 \tag{27}$$

So we can write:

$$Re[\lambda_{i}(A_{Sys} - B_{Sys}.K_{\alpha})] < -\alpha \tag{28}$$

In the other hand, all of eigenvalues of (ASys-BSys.Kα) are located on the left hand-side of -α in S plane. So the control signal U, which minimizes the performance index in (21), can create a closed lope system with determined degree of stability.

### 3.1.1 Numeral sample

This subsection gives a dimensional example in which the controllable system is to be designed in according to the proposed method. Suppose the nominal plant used for design of the controller is:

$$\frac{d}{dt}X = \begin{bmatrix} 0 & 1 \\ 0 & 0 \end{bmatrix}.X + \begin{bmatrix} 1 & 0 \\ 0 & 1 \end{bmatrix}.U \tag{29}$$

Where $X=[x_1(t) \; x_2(t)]^T$ and $U=[\mu_1(t) \; \mu_2(t)]^T$. We want to design an optimal controller in which the closed-loop system achieves to a favourite prescribed degree of stability. In this case, we chose $\alpha=2$. Suppose that

$$Q = \begin{bmatrix} 1 & 1 \\ 1 & 1 \end{bmatrix} \quad , \quad R = \begin{bmatrix} 1 & 0 \\ 0 & 1 \end{bmatrix} \tag{30}$$

The solution of Riccati equation in this case is equal to:

$$P_\alpha = \begin{bmatrix} 3.9316 & 1.1265 \\ 1.1265 & 4.4462 \end{bmatrix} \tag{31}$$

To obtain the feedback control law in terms of system states, we can combine (23) with (29 – 31). Finally the optimal controller is given by:

$$U_\alpha(t) = -R^{-1}.B^T.P_\alpha.x(t) = \begin{bmatrix} -3.9316 & -1.1265 \\ -1.1265 & -4.4462 \end{bmatrix}.x(t) \tag{32}$$

And then the state matrix of optimal closed loop system $(A_\alpha)$ is obtained by:

$$A_\alpha = A + 2I_n - B.K_\alpha = \begin{bmatrix} -3.9316 & -0.1265 \\ -1.1265 & -4.4462 \end{bmatrix} \tag{33}$$

So the poles of optimal closed loop system are located in -3.7321and -4.6458. It can be seen that both of them shift to the left hand-side of $-\alpha$ ($\alpha=2$) in S plane.

### 3.2 Reduced order observer

The Luenberger reduced-order observer is used as a linear observer in this paper. The block diagram of this reduced-order observer is shown in Fig. 3. For the controllable and observable system that is defined by (15), there is an observer structure with size of (n-1). The size of state vector is n and output vector is l. The dynamic system of Luenberger reduced-order observer with state vector of z(t), is given by:

$$\Delta z(t) = L.\Delta x_{Total}(t) \tag{34}$$

$$\dot{z}(t) = D.z(t) + T.y_{Total}(t) + R.u_{Total}(t) \tag{35}$$

To determine L, T and R is basic goal in reduced-order observer. In this method, the estimated state vector $\Delta \hat{X}_{Total}(t)$ includes two parts. First one will obtain by measuring $\Delta y_{Total}(t)$ and the other one will obtain by estimating $\Delta z(t)$ from (34). We can take:

$$\begin{bmatrix} \Delta y_{Total}(t) \\ \Delta z(t) \end{bmatrix} = \begin{bmatrix} C_{Total} \\ L \end{bmatrix}.\Delta \hat{X}_{Total}(t) \tag{36}$$

By assumption full rank $[C_{Total}^T \; L^T]^T$, we can get:

$$\Delta \hat{X}_{Total}(t) = \begin{bmatrix} C_{Total} \\ L \end{bmatrix}^{-1} \cdot \begin{bmatrix} \Delta y_{Total}(t) \\ \Delta z(t) \end{bmatrix} \tag{37}$$

By definition:

$$\begin{bmatrix} C_{Total} \\ L \end{bmatrix}^{-1} = \begin{bmatrix} F_1 & F_2 \end{bmatrix} \tag{38}$$

We get:

$$\Delta \hat{X}_{Total}(t) = F_1 . \Delta y_{Total}(t) + F_2 . \Delta z(t) \tag{39}$$

Where:

$$F_1 . C_{Total} + F_2 . L = I_n \tag{40}$$

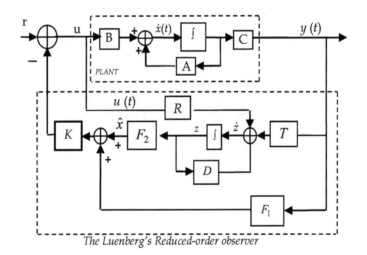

*The Luenberg's Reduced-order observer*

Fig. 4. Shematic Diagram of State Feedback Using Luenberger Reduced Order Observer

Using estimated state variables, the state feedback control law is given by:

$$\Delta U_{Total}(t) = -K.\Delta \hat{X}_{Total}(t) = K.F_1.C_{Total}.\Delta X_{Total}(t) - K.F_2.\Delta z(t) \tag{41}$$

By assumption $R = L.B_{Total}$ in (35), descriptive equations of closed loop control system with reduced-order observer are:

$$\begin{bmatrix} \Delta \dot{X}_{Total}(t) \\ \Delta \dot{z}(t) \end{bmatrix} =$$

$$\begin{bmatrix} A_{Total} - B_{Total}.F_1.C_{Total} & -B_{Total}.K.F_2 \\ T.C_{Total} - L.B_{Total}.K.F_1.C_{Total} & D - L.B_{Total}.K.F_2 \end{bmatrix} . \begin{bmatrix} \Delta X_{Total}(t) \\ \Delta z(t) \end{bmatrix} \tag{42}$$

Dynamic error between linear combination of states of $L.\Delta X_{Total}(t)$ system and observer $\Delta z(t)$ is defined as:

$$\dot{e}(t) = \Delta \dot{z}(t) - L.\Delta \dot{X}_{Total}(t) \tag{43}$$

Combine (35) and (36), we get:

$$\begin{bmatrix} \Delta \dot{X}_{Total}(t) \\ \dot{e} \end{bmatrix} = \begin{bmatrix} A_{Total} - B_{Total}.K & -B_{Total}.K.F_2 \\ 0 & D \end{bmatrix}.\begin{bmatrix} \Delta X_{Total}(t) \\ e \end{bmatrix} \tag{44}$$

For stability of the observer dynamic system, the eigenvalues of D must lie in the left hand-side of s plane. By choosing D, we can calculate L, T and R (Luenberger, 1971).

### 3.2.1 Numeral sample

The longitudinal equations of an aircraft are presented in steady state format (Rynaski, 1982):

$$\frac{d}{dt}\begin{bmatrix} \Delta v(t) \\ \alpha(t) \\ \theta(t) \\ q(t) \end{bmatrix} = \begin{bmatrix} \alpha_{11} & a_{12} & 0 & -g \\ a_{21} & a_{22} & 1 & 0 \\ a_{31} & a_{32} & a_{33} & 0 \\ 0 & 0 & 1 & 0 \end{bmatrix}x(t) - \begin{bmatrix} b_1 \\ b_2 \\ b_3 \\ 0 \end{bmatrix}\delta_e(t) \tag{45}$$

Where $\Delta v$ is variation of velocity, $\alpha$ is angle of attack, $\theta$ is pitch angle, $\delta_e$ is elevator angle and q is pitch rate. This equation in special state is presented by:

$$\frac{d}{dt}X(t) = \begin{bmatrix} -0.0507 & -3.861 & 0 & -9.8 \\ -0.00117 & -0.5164 & 1 & 0 \\ -0.000129 & 1.4168 & -0.4932 & 0 \\ 0 & 0 & 1 & 0 \end{bmatrix}X(t) - \begin{bmatrix} 0 \\ -0.0717 \\ -1.645 \\ 0 \end{bmatrix}\delta_e(t) \tag{46}$$

$$y = \begin{bmatrix} 1 & 0 & 0 & 0 \\ 0 & 1 & 0 & 0 \end{bmatrix}X(t) \tag{47}$$

So $\Delta v$ and $\alpha$ are measurable state variables. Therefor we can get final result by MATLAB simulations.

## 4. Simulation results

Eigenvalue analysis is a fast and well-suited technique for defining behavioral trends in a system that can provide an immediate stability test. The real parts of the eigenvalue represent the damping mode of vibration, a positive value indicating instability, while the imaginary parts denote the damped natural frequency of oscillation.

As mentioned earlier, the system considered here is the IEEE second benchmark model. It is assumed that the fixed capacitive reactance $(X_C)$ is 81.62% of the reactance of the transmission line $(X_{L1}=0.48$ P.u).

The simulation studies of IEEE-SBM carried out on MATLAB platform is discussed here. The following cases are considered for the analysis.

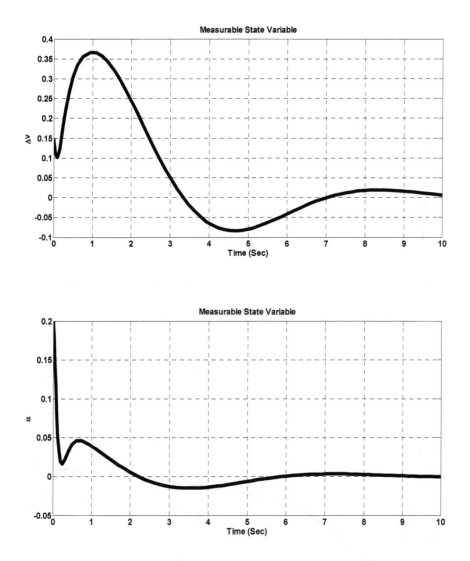

Fig. 5. Variation of Measurable State Variables

## 4.1 Without controller

In the first Case, second benchmark model is simulated without any controller in initial conditions. Fig. 7 shows the variation of torque of the rotating mechanical system of the T-G set. It can be seen that variations of torque of the mechanical system are severe unstable and then power system tends to approach to the SSR conditions.

The study is carried out with heavily loaded synchronous generator of $P_G$=0.9 p.u, $Q_G$=0.43 p.u and $V_t$=1.138 p.u. Figure. 8 shows the variation of real and imaginary parts of the eigenvalues of $A_{Sys}$ with the compensation factor $\mu_C$=$X_C$/$X_{L1}$. It can be observed that the first torsional mode is the most unstable modes at $\mu_C$=81.62%. The unstable range of variation of torsional modes has been illustrated in Fig. 8.

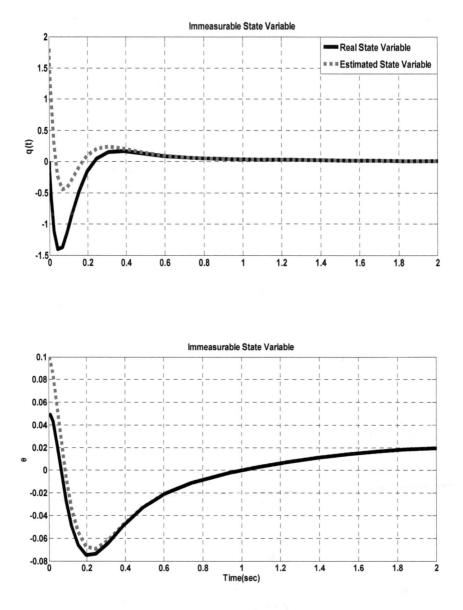

Fig. 6. Variation of Immeasurable State Variables

## 4.2 With proposed controller: Extended optimal control

For presentation of the first proposed controller, power system is simulated by extended optimal control with using (20-24). Proposed method is carried out on second benchmark simultaneously. The obtained results have been compared with prevalent optimal control in (17-20) by Fig. 9-(a). It is observed that the proposed method has created more suitable damping for first torsional mode of second benchmark model than prevalent method. As similar way, this proposed method has suitable results on second torsional mode. Fig. 9-(b) clearly illustrates this point.

Fig. 10 illustrates variation of torque of the mechanical system in T-G set. It can be observed that the proposed method has more effect on the output of power system than prevalent method.

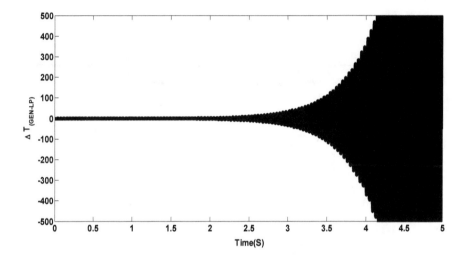

Fig. 7. Variation of torque of generator – low pressure turbine in the T-G set for $\mu_C$=81.62% without any controller.

### 4.3 With proposed controller: Reduced order observer

In order to have a complete research, optimal full state feedback control is designed and the results are compared with reduced-order method. Some parameters, such as $\Delta i_{kd}$ and $\Delta i_{kq}$, are not physical variables. $\Delta V_{Cd}$ and $\Delta V_{Cq}$ are transmission line parameters that they are not accessible. So let us define:

$$\Delta \hat{X}_{Sys}{}^{T} = [\Delta i_{kd} \quad \Delta i_{kq} \quad \Delta V_{Cd} \quad \Delta V_{Cq}] \tag{48}$$

$$\Delta y_{Sys}{}^{T} = [\Delta T_{EXC-GEN} \quad \Delta T_{GEN-LP} \quad \Delta T_{LP-HP}] \tag{49}$$

Where $\Delta y_{Sys}{}^{T}$ is used to obtain variation of torque of the rotating mechanical system of the T-G set . Full order observer estimates all the states in a system, regardless whether they are measurable or immeasurable. When some of the state variables are measurable using a reduced-order observer is so better.

In this scenario, proposed method is carried out on second benchmark model simultaneously. The obtained results have been illustrates in Fig. 11. It is observed that the reduced-order method has created a suitable estimation from immeasurable variables that are introduced in (48).

Fig. 12 shows variation of torque of the mechanical system in T-G set. It can be observed that the proposed method has small effect on the output of power system

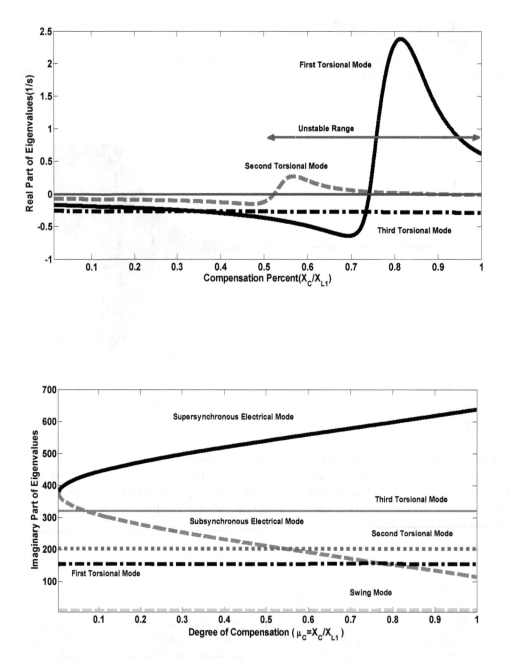

Fig. 8. Variation of real and imaginary parts of eigenvalues as a function of $\mu_C$

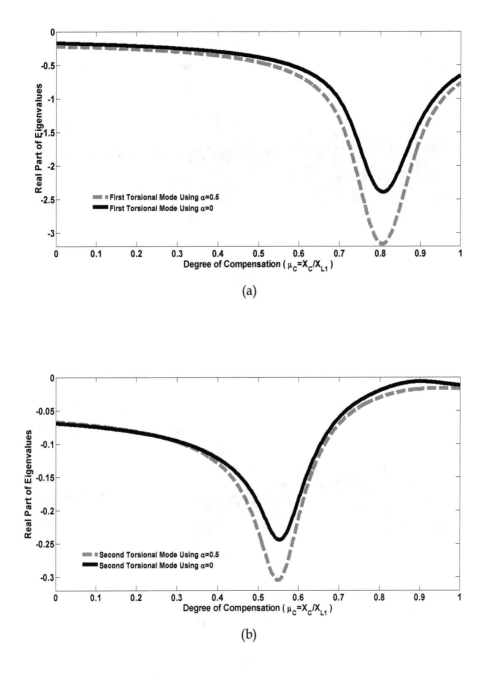

Fig. 9. Variation of first torsional mode (a) and second mode (b) in IEEE second benchmark to degree of compensation: Dotted (proposed controller), Solid (prevalent controller).

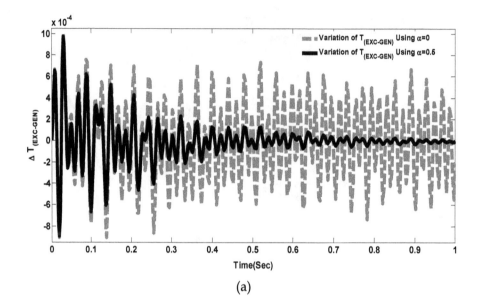

(a)

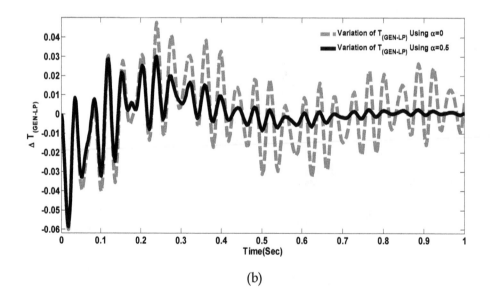

(b)

Fig. 10. Variation of torque of exciter-generator (a) and generator-low pressure (b):
(proposed controller), Dotted (prevalent controller).

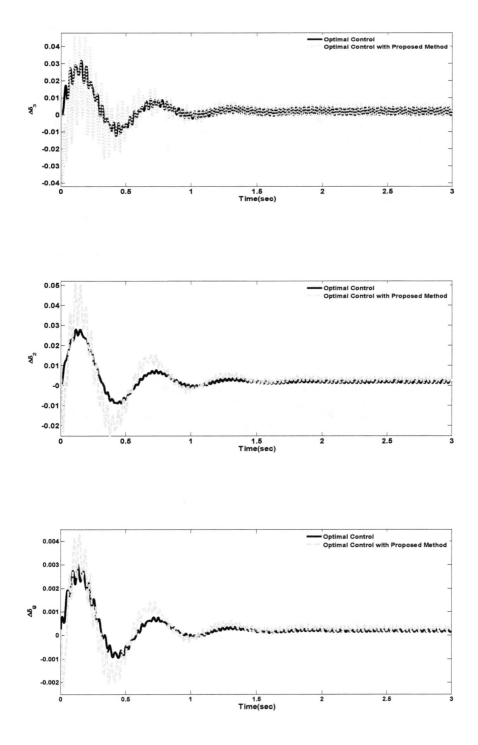

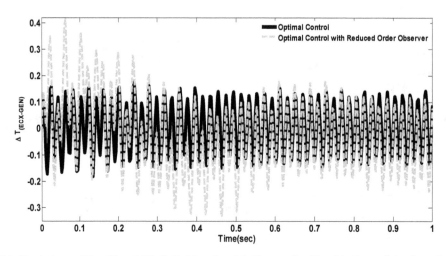

Fig. 11. Variation of δg, δ2and δ3: Solid (optimal full state feedback), Dotted (reduced-order observer control).

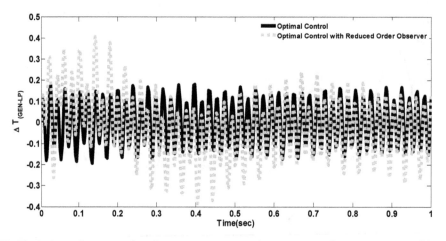

Fig. 12. Variation of torque of exciter – generator and generator – low pressure turbine in the T-G set: (reduced-order observer control), Solid (optimal full state feedback).

## 5. Conclusion

Fixed capacitors have long been used to increase the steady state power transfer capabilities of transmission lines. A major concern associated with fixed series capacitor is the sub-synchronous resonance (SSR) phenomenon which arises as a result of the interaction between the compensated transmission line and turbine-generator shaft. This results in excessively high oscillatory torque on machine shaft causing their fatigue and damage.

This chapter presents two analytical methods useful in the study of small-signal analysis of SSR, establishes a linearized model for the power system, and performs the analysis of the SSR using the eigenvalue technique. It is believed that by studying the small-signal stability of the power system, the engineer will be able to find countermeasures to damp all sub-synchronous torsional oscillations.

The first strategy is proposed, based on shifting eigenvalues of the state matrix of system to the left hand-side of s plane, for damping all sub-synchronous torsional oscillations. The proposed method is applied to The IEEE Second Benchmark system for SSR studies and the results are verified based on comparison with those obtained from digital computer simulation by MATLAB. Analysis reveals that the proposed technique gives more appropriate results than prevalent optimal controller. In the practical environment (real world), access to all of the state variables of system is limited and measuring all of them is also impossible. So when we have fewer sensors available than the number of states or it may be undesirable, expensive, or impossible to directly measure all of the states, using a reduced-order observer is proposed. Therefore in this chapter, another novel approach is introduced by using optimal state feedback, based on the Reduced – order observer structure. Analysis reveals that the proposed technique gives good results. It can be concluded that the application of reduced-order observer controller to mitigate SSR in power system will be provided a practical viewpoint. Also this method can be used in a large power system as a local estimator.

## 6. Acknowledgement

The authors would like to express their sincere gratitude and appreciations to the Islamic Azad University, Gonabad Branch, as collegiate assistances in all stages of this research.

## 7. Appendix (Nomenclature)

| | |
|---|---|
| $\Delta X_{Gen}$ | State Vector for Generator System Model |
| $A_G$ | State Matrix for Generator System Model |
| $B_{Gi}$ | $i$th Input Matrix for Generator System Model |
| $\Delta y_{Gen}$ | Output Vector for Generator System Model |
| $\Delta U_{Gen}$ | Input Vector for Generator System Model |
| $\Delta i_{fd}$ | Variation of Field Winding Current |
| $\Delta i_d, \Delta i_q$ | Variation of Stator Currents in the d-q Reference Frame |
| $\Delta i_{kd}, \Delta i_{kq}$ | Variation of Damping Winding Current in the d-q Reference Frame |
| $\Delta \delta_g$ | Variation of Generator Angle |
| $\Delta \omega_g$ | Variation of Angular Velocity of Generator |
| $\Delta X_{Mech}$ | State Vector for Mechanical System Model |
| $A_M$ | State Matrix for Mechanical System Model |
| $B_{Mi}$ | $i$th Input Matrix for Mechanical System Model |
| $\Delta y_{Mech}$ | Output Vector for Mechanical System Model |
| $\Delta U_{Mech}$ | Input Vector for Mechanical System Model |
| $\Delta T_m$ | Variation of Mechanical Torque |
| $\Delta T_e$ | Variation of Electrical Torque |
| $\Delta X_{Line}$ | State Vector for Transmission Line System |
| $A_{Line}$ | State Matrix for Transmission Line System |
| $B_{Line}$ | Input Matrix for Transmission Line System |
| $\Delta U_{Line}$ | Input Vector for Transmission Line System |
| $\Delta X_{Sys}$ | State Vector for Combined Power System Model |

$A_{Sys}$      State Matrix for Combined Power System Model
$B_{Sys}$      Input Matrix for Combined Power System Model
$\Delta U_{Sys}$      Input Vector for Combined Power System Model
$\Delta E$      Variation of Field Voltage
$J$      Performance Index
$K$      Gain Feedback Vector in Linear Optimal Control
$K_\alpha$      Gain Feedback Vector in Linear Optimal Control with Determined Degree of Stability
$P$      Solution of Riccati Equation in Linear Optimal Control
$P_\alpha$      Solution of Riccati Equation in Linear Optimal Control with Determined Degree of Stability

## 8. References

Walker D.N.; Bowler C.L., Jackson R.L. & Hodges D.A. (1975). Results of sub-synchronous resonance test at Mohave, *IEEE on Power Apparatus and Systems*, pp. 1878–1889

Hall M.C. & Hodges D.A. (1976). Experience with 500 kV Subsynchronous Resonance and Resulting Turbine-Generator Shaft Damage at Mohave Generating Station, *IEEE Publication*, No.CH1066-0-PWR, pp.22-29

IEEE Committee report. (1992). Reader's guide to Subsynchronous Resonance. *IEEE on Power Apparatus and Systems*, pp. 150-157

IEEE Torsional Issues Working Group. (1997). Fourth supplement to a bibliography for study of sub-synchronous resonance between rotating machines and power systems, *IEEE on Power Apparatus and Systems*, pp. 1276–1282

Anderson P.M. ; Agrawal B. L. & Van Ness J.E. (1990). *Sub-synchronous Resonance in Power System*, IEEE Press, New York

Begamudre R.D. (1997). *Extra High Voltage A.C Transmission Engineering*, New Age International (P) Limited, New Delhi

Gross G,; Imparato C. F., & Look P. M. (1982). A tool for comprehensive analysis of power system dynamic stability, *IEEE on Power Apparatus and Systems*, pp. 226–234, Jan

Parniani M. & Iravani M. R. (1995). Computer analysis of small-signal stability of power systems including network dynamics, *Proc. Inst. Elect. Eng. Gener. Transmiss. Distrib.*, pp. 613–617, Nov

Harb A.M. & Widyan M.S. (2004). Chaos and bifurcation control of SSR in the IEEE second benchmark model, *Chaos, Solitons and Fractals*, pp. 537-552. Dec

Zhu W.; Mohler R.R., Spce R., Mittelstadt W.A. & Mrartukulam D. (1996). Hopf Bifurcation in a SMIB Power System with SSR, *IEEE Transaction on Power System*, pp. 1579-1584

Patel R.V. & Munro N. (1976). *Multivariable System Theory and Design*, Pergamon Press

Khaki Sedigh A. (2003). *Modern control systems*, university of Tehran, Press 2235

Ogata K. (1990). *Modern Control Engineering*, Prentice–Hall

Friedland B. (1989). *Control System Design : An Introduction to State – Space Methods*, Mc Graw–Hill

Kwakernaak H. & Sivan R. (1972). *Linear Optimal Control Systems*, Wiley- Intersciences

Luenberger D.G. (1971). An Introduction to Observers, *IEEE Trans On Automatic Control*, pp 596-602 , Dec

Rynaski E.J. (1982). Flight control synthesis using robust output observers, *Guidance and Control Conference*, pp. 825-831, San Diego

# 8

# Automatic Guided Vehicle Simulation in MATLAB by using Genetic Algorithm

Anibal Azevedo
*State University of São Paulo*
*Brazil*

## 1. Introduction

The prodigious advances in robotics in recent times made the use of robots more present in modern society. One important advance that requires special attention is the development of an unmanned aerial vehicle (UAV), which allows an aircraft to fly without having a human crew on board, although the UAVs still need to be controlled by a pilot or a navigator.

Today's UAVs often combine remote control and computerized automation in a manner that built-in control and/or guidance systems perform deeds like speed control and flight-path stabilization. In this sense, existing UAVs are not truly autonomous, mostly because air-vehicle autonomy field is a recently emerging field, and this could be a bottleneck for future UAV development.

It could be said that the ultimate goal in the autonomy technology development is to replace human pilots by altering machines decisions in order to make decisions like humans do. For this purpose, several tools related with artificial intelligence could be employed such as expert systems, neural networks, machine learning and natural language processing (HAYKIN, 2009). Nowadays, the field of autonomy has mostly been following a bottom-up approach, such as hierarchical control systems (SHIM, 2000).

One interesting methodology from the hierarchical control systems approach is the subsumption architecture that decomposes complicated intelligent behavior into many "simple" behavior modules, which are organized into layers. Each layer implements a particular goal and higher layers are increasingly more abstract. The decisions are not taken by a superior layer, but by listening to the information that are triggered by sensory inputs (lowest layer). This methodology allows the use of **reinforcement learning** to acquire behavior with the information that comes with experience.

Inspired by old behaviorist psychology, **reinforcement learning** (RL) concerned with how an *agent* ought to take *actions* in an *environment,* so as to maximize some notion of cumulative *reward*. Reinforcement learning differs from standard supervised learning in that correct input/output pairs which are never presented (HAYKIN, 2009). Furthermore, there is a focus on an on-line performance, which involves finding a balance between exploration (of uncharted territory) and exploitation (of current knowledge). The reinforcement learning has been applied successfully to various problems, including robot control, elevator scheduling, telecommunications, backgammon and chess (SHIM, 2000).

**Genetic algorithms** (GAs) are developed in order to emulate the process of genetic evolution found in nature, and then perform artificial evolution. They were developed by John Holland [11] in the early 70s, and have been successfully applied to numerous large and complex search space problems ever since (MICHALEWICZ, 1996).

In nature, organisms have certain characteristics that affect their ability to survive and reproduce. These characteristics are contained in their genes. Natural selection ensures that genes from a strong individual are presented in greater numbers in the next generation, rather than those from a weak individual. Over a number of generations, the fittest individuals, in the environment in which they live, have the highest probability of survival and tend to increase in number; while the less fit individuals tend to die out. This is the Darwin's principle of the survival-of-the-fittest and constitutes the basic idea behind the GAs. In order to perform computational tests on how the reinforcement learning could cope with genetic algorithms to provide good rules for the navigation of an automatic vehicle (STAFYLOPATIS, 1998), a program that emulates a navigation environment was developed in a Matlab. This Chapter will describe how the ideas developed by (STAFYLOPATIS, 1998) could be employed to study the integration of the GAs and RL to produce rules for automatic guided vehicles searching for a better performance. The main contributions of this Chapter are: the vehicle, its sensors, and also the environment for training are different from the ones presented in (STAFYLOPATIS, 1998); a new equation for the reinforcement learning was proposed; the influence of the parameters that control the production of automatic generation of rules for vehicle control navigation are also tested. Sections 2 and 3 describe how the navigation decision rules could be encoded in a vector. Sections 4 and 5 show how the reinforcement learning and genetic algorithm uses the encoding of Section 2 to cope with the production of vehicle control navigation rules. In Section 6, some results are presented and finally, in Section 7, some conclusions and future works are given. All the proposed approach had been implemented in a Matlab.

## 2. Navigation problem representation

The navigation problem could be defined as how a vehicle could be guided through an ambient with many obstacles and barriers using the information available from the information given by the vehicle radar, as it can be seen in Figure 1.

Figure 1 also details how the vehicle radar works. The radar has 9 sectors in order to better acquire the information of how near is a vehicle to an object. The information of existence of an obstacle for each section is stored in vector **v** of nine positions using the following rule: if one object is in one section, then value 1 is attributed. If not, then value 0 is attributed. Figure 2 details a situation where there are obstacles in sector 1 and 2 and the correspondent representation by a vector. Once the objects had been detected by the radar, it is necessary to implement an appropriate action which could be one of the following three: turn 15° degrees to the left, keep the trajectory or turn 15° degrees to the right. For the sake of clarity, but without loss of generality, velocity $v_c$ of the vehicle was assumed to be constant. With these three possible decisions, a new concept could be created, which is a rule. A **rule** is a vector that combines the one that describes the situation for the vehicle in terms of obstacles, like the one presented in Figure 2, and a new component that decides the vehicle action in turn to avoid a set of obstacles: 0 – turn left, 1 – keep the trajectory, 2 – turn right. An example of a rule for the situation showed in Figure 2 is described in Figure 3.

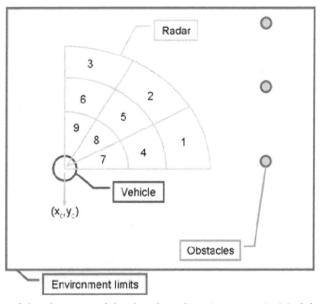

Fig. 1. Description of the elements of the developed environment in Matlab.

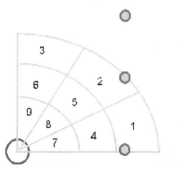

Situation  1  1  0  0  0  0  0  0  0

Vector
Index    0  1  2  3  4  5  6  7  8

Sector   1  2  3  4  5  6  7  8  9

Fig. 2. The correspondence between the obstacles detected by a radar vehicle and a vector with binary information.

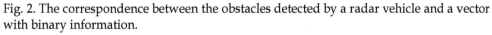

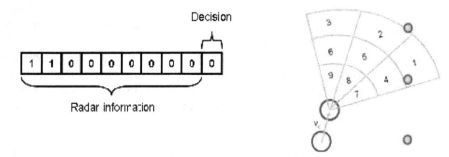

Fig. 3. The correspondence between a rule and decision taken by the vehicle.

## 3. Navigation control

The previous binary representation scheme has a main drawback since before every vehicle movement decision demanded to store $3 \times 2^9 = 1536$ vectors (rules) to precisely describe which action should be performed for each scenario detected by the radar information. This implies that before every step performed by the vehicle, a control action should compare, in the worst case, 1536 vectors in order to find the proper decision to be taken. This scheme is illustrated in Fig. 4.

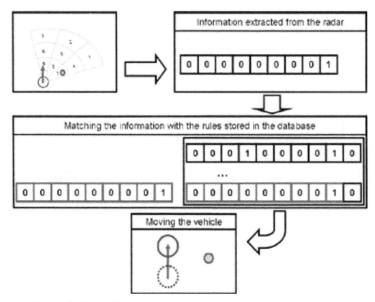

Fig. 4. Decision scheme followed by the automatic guided vehicle.

This computational work could be avoided by introducing a new symbol 2, for the radar information section, which means that "it could have or not an obstacle at this sector". The advantage of the definition of this new symbol is that it will help to reduce the number of necessary information to be kept by the vehicle control. The disadvantage of using this new symbol is that it could group different situations where the decisions should not be the same. Fig. 5 gives an example of this situation.

The example showed in Fig. 5 emphasizes the importance to construct the rules in a manner that the parts of the rule which do not affect the decision should be numbered as 2, and the other parts that have a great influence in the final behavior of the vehicle should be numbered as most specific as possible or, in other words, with the numbers 0 or 1. Fig. 6 illustrates the application of this concept.

Fig. 6 also gives a guideline procedure for another problem that could emerge, which is the appearance of more than one rule that matches the current radar information when symbol 2 is used. One criteria that will be adopted is to select the rules which match the environment situation, but with as much specific information as possible. For this purpose Eq. (1) will be adopted.

$$S_i = \frac{(n - k_i)}{n} \qquad i = 1, \cdots, R \tag{1}$$

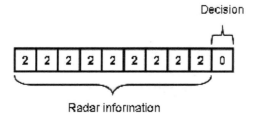

(A) The rule that match the acquired information obtained by the vehicle radar.

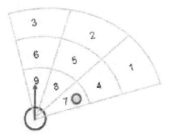

(B) Employing the rule – Step 1, Situation 1.          (C) Employing the rule – Step 2, Situation 1.

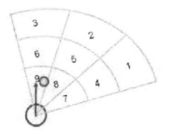

(D) Employing the rule – Step 1, Situation 2.          (E) Employing the rule – Step 2, Situation 2.

Fig. 5. Employing the same rule in different scenarios and the corresponding consequences.

Where: $S_i$ is the measure of how specific is rule $i$, $n$ is the number of sectors the vehicle radar has, $k_i$ is the number of digits marked with value 2, and $R$ is the number of rules. Thus $S_i \in [0,1]$ and note that variable $S_i$ becomes 0 when every rule digit related with the sector state is marked with value 2 (lowest specificity) and $S_i$ becomes 1, if value 2 does not appear in rule $i$ (highest specificity).

This discussion also shows that the efficiently vehicle navigation control depends on more than one rule, and this leads to the objective of finding the set of rules that could properly control the vehicle. To achieve this objective the next sections will discuss concepts on how to automatically construct this set of rules by employing computational tools such as Reinforcement Learning and Genetic Algorithm.

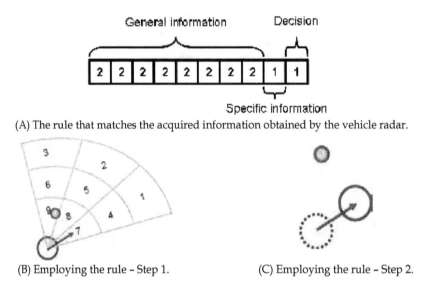

(A) The rule that matches the acquired information obtained by the vehicle radar.

(B) Employing the rule – Step 1.                          (C) Employing the rule – Step 2.

Fig. 6. Employing a rule that combines general (number 2) with specific (number 0 or 1) knowledge.

## 4. Reinforcement learning

The developed program starts with rules totally generated at random and which quality, in terms of helping the vehicle navigation control, is unknown. Then, it was created a system that evaluates the quality of rules used through the vehicle navigation performed by a computational simulation. As it was seen before, during the navigation, more than one rule could be necessary and the system must punish or reward not only a rule, but also all rules employed that lead the vehicle to a collision or to avoid an obstacle, respectively. One difficulty related with this scheme is to know the correct contribution weight of each rule in a possible vehicle collision. It is also difficult to estimate the reward value for a set of rules that helps the vehicle to avoid an immediate collision, because this set could put the vehicle in a trajectory that leads to a future collision with another obstacle. Besides those problems, the initial credit received for each rule is modified by Eq. (2), if there are no collisions on the current vehicle step $t$.

$$C_{it+1} = C_{it} + kS_{it}C_{it} \qquad (2)$$

Where: $i$ is the rule that perfectly matches with the current vehicle situation; in order words, rule $i$ is the one which $S_{it}$ value is as greater as possible at the step $t$ of the vehicle, $C_{it}$ is the credit value assigned for each rule $i$ at the step $t$, and $k$ is a constant that indicates the learning rate.

When the vehicle crashes at some obstacle a punishment must be applied to the rules mostly related to the event. To avoid the storage of large amounts of historical record about what rules where used, and to punish just the rules mostly related to the collision, only the last three employed rules will have their credit updated by Eq. (3).

$$C_{it+1} = C_{it} / 2 \qquad (3)$$

Eq. (3) was considered by (STAFYLOPATIS, 1998), but it represents that the rules that lead the vehicle to perform more than 200 steps without a collision will suffer the same punishment as the rules that conduct the vehicle to perform only 20 steps without a collision, for example. To avoid this problem, this book Chapter proposes, for the first time, a punishment formula which is a function on the number of steps as showed in Eq. (4) and (5).

$$C_{it+1} = k_p C_{it} \qquad (4)$$

$$k_p = \frac{1}{1 + \exp^{-(ns+1)/np}} \qquad (5)$$

The behavior of Eq. (5) for $np = 20$ is showed in Fig. 7.

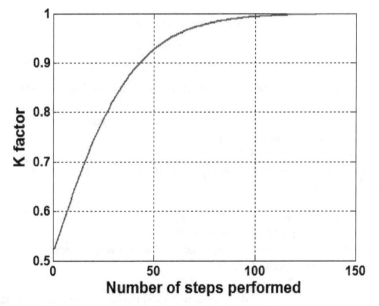

Fig. 7. Behavior of the parameter penalty factor $k_p$ for $np = 20$.

The purpose of Eq. (4) and (5) is to punish the rules according with total number of steps ($ns$) performed by the vehicle before the collision, and it was reflected in a different credit actualization. For example, if the three last used rules lead to only 25 steps without a collision, then the new credit value will be 75% of the old credit value (see Fig. 7). The system proposed by Eq. (2) and (3) or (4) and (5) starts by assigning the same value, same credit, for every rule. Then, the rules performance is evaluated in an environment, created in Matlab, where the vehicle control is simulated, and during the simulation, after each step performed by the vehicle, it is verified whether there is a collision (then the Eq. (3) or Eq. (4) and (5) is used) or not (then the Eq. (2) is used) following the scheme described in Fig. 8.

```
function EvaluateRules (MR)
  % Initial learning rate for each step without a collision.
  k = 0.2;
  collision = 0;
  % Initial credit assignment for all rules.
  for i=1:R
    C(i) = 0.2;
  end
  % Starts the initial conditions of the environment and the objects that appear in the radar
  % vehicle.
  [vE, E]=Starts();
  ns = 0;          % Number of steps.
  Records = [ ];   % Stores the index of the rules used.
  % Loop that evaluates the rules inside MR while there is no collision.
  While (collision == 0)
    % Determining the rules that matches with the actual environment.
    [vm]=Match(MR, vE);
    % Verifying the measure of how specific each matching rule is (use Equation (1)).
    MRE = MR(vm,:);
    [sm]=Specific(MRE);
    [value, indexsmax] = max(sm);
    % Using the matching rule with maximum Sit value and verifying the consequences.
    VRM = MRE(indexsmax,:);
    % Update the vehicle in environment by using the selected rule. It also returns if
    % a collision happened (collision == 1).
    [collision, vE, E]= Simulate(VRM, E) ;
    % There is no collision with the rule application, then update the credit according
    % with the equation (2).
    If (collision == 0)
      % Update the credit of the rule applied.
      C(vm(indexsmax)) = (1 + k*sm(indexsmax))*C(vm(indexsmax))
      % Store the index of what rule was applied in a certain vehicle step.
      Records = [Records vm(indexsmax)];
      ns = ns + 1;
    % There is a collision after the rule application then update the credit of the last three
    % rules used according with equation (3) or (4) and (5).
    else
      indexp = Records(end-2:end);
      % Or use: kp = 1/(exp(-(ns+1)/20)+1); C(indexp)=C(indexp)*kp;
      C(indexp) = C(indexp)/2;
      ns = 0;
    end
  end
end
```

Fig. 8. The scheme to reward or to punish the rules during a simulation in the environment.

The functions and symbols used in Fig. 8 are defined as follows:

$C$         - Vector containing the credit associated to a set of rules.

$E$         - Matrix with all data about the current simulation environment state.

$vE$        - Vector with information extracted from the radar, as show in Figure 2.

$MR$        - Matrix with a set of rules. Each matrix line represents one rule.

$Vm$        - The vector with the index of the rules that matches the information extracted from vehicle radar (contained in vE).

$MRE$       - Matrix with only the rules that match the information extracted from vehicle radar.

*Specific*  - This function evaluates all the rules contained in matrix MRE according with the equation Eq. (1) in order to measure how specific is each rule.

$sm$        - Vector with all rules specific measure.

*max*       - This function determines which specific measure is the biggest (value) and the corresponding line of MRE (indexsmax).

$VRM$       - Vector with the rule selection to be used in current environment state.

*Simulate*  - This function simulates the vehicle through environment using VRM rules.

*collision* - Variable that indicates whether a vehicle collides (equals to 1) or not (equals to 0).

*indexsmax* - Index of the rule with maximum matching value.

*Records*   - Vector that stores the index of the rules used in a certain vehicle step through the environment.

*indexp*    - Vector with the index of the last three rules applied before the vehicle collision and which will be punished by a credit decreasing.

$ns$        - Number of steps without a collision.

$kp$        - Penalization factor applied in the credit of the last three rules when a collision occurs.

The system showed in Fig. 8 tries to establish a reward and a punishment scheme among the rules and their impact in the vehicle navigation through equations that monitor the performance, although the random generation of rules can also produce sets of rules which lead to a bad control navigation performance. After a pre-specified number of collisions, the set of rules could be changed by a new randomly generated set of rules formed with the insertion of new rules, the exclusion of rules with a bad performance (low credit value), and the maintenance of rules that help to avoid the obstacles (high credit value). To perform the formation of this new set of rules, a genetic algorithm was coupled to the credit evaluation scheme described in Fig. 8, as could be seen in Fig.9.

The complete description of the Genetic Algorithm developed is described in Section 5.

## 5. Genetic algorithm

The genetic algorithm keeps a population of individuals, represented by: $A(t) = \{A_1^t, ..., A_n^t\}$ for each generation (iteration) $t$, and each individual represents a rule to guide the vehicle through the environment and to avoid the obstacles. In the computational implementation adopted here, the population is stored in a matrix $A(t)$ and each column $A_i^t$ represents a rule. Each rule $A_i^t$ is evaluated according to the number of vehicle movements without a collision, and then, the *fitness*, a measure of how this individual is sucessful for the problem, is calculated. The *fitness* is calculated for the entire population and is based on this new population that combines the most capable individuals that will form generation $t+1$.

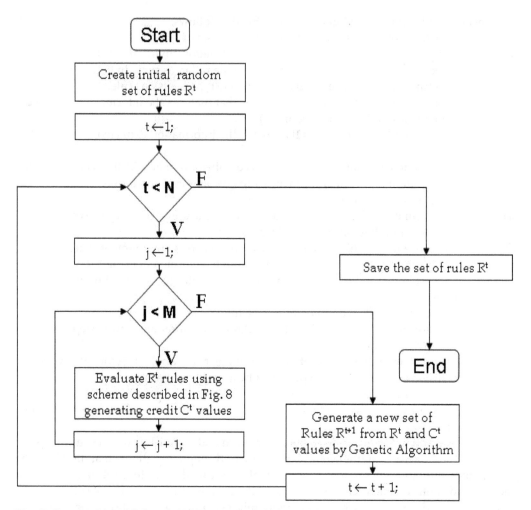

Fig. 9. Composition of the credit scheme and Genetic Algorithm.

During the formation of the new population, some individuals from generation $t$ are submitted to a transformation process by genetic operators in order to form new rules. These transformations include unary operators $m_i$ (mutation), that allow the creation of new rules by small changes in the individual attributes ($m_i$: $A_i \rightarrow A_i$), and superior order transformation $c_j$ (crossover) that produces new individuals by combining one or more individuals ($c_j$: $A_j \times \ldots \times A_k \rightarrow A_j$). This process is carried out until a previous specified maximum number of generations is reached (MICHALEWICZ, 1996).

During the vehicle navigation along the environment different situations may occur, and each one of them could require a proper action implying the necessity of using more than just one rule. For this reason, the **reinforcement leaning** is acopled with a population produced by the genetic algorithm and, when a collision occurs, a proper number of rules used before the collision are penalized (STAFYLOPATIS, 1998). This system consists in firstly give the same value for each rule of the population, and then, evaluate each rule according to the number of times it was used until a collision happened.

The implementation details adopted for the Genetic Algorithm are as follows:

**(A) Data structure codification for each individual:**

Each genetic algorithm individual is associated to a rule by using a vector $v$ with 10 elements and values inside the elements 1 to 9, corresponding to the presence (value 1) or not (value 0) or whatever (value 2) in the corresponding vehicle radar section, respectively (see Fig. 3). The value inside the position 10 indicates the decision that has to be made: 0 - turn 15° degrees to the left, 1 – keep the trajectory and 2 - turn 15° degrees to the right. The set of rules is stored in a matrix and each column represents a rule, as showed in Fig. 10. In the Example showed in Fig. 10, the first column contains a vector $\mathbf{v_1}$ with 10 positions which each position meaning explained in Fig. 2.

| Sector | Rule 1 | Rule 2 |
|--------|--------|--------|
| 1 | 0 | 2 |
| 2 | 0 | 2 |
| 3 | 1 | 1 |
| 4 | 2 | 0 |
| 5 | 2 | 2 |
| 6 | 2 | 2 |
| 7 | 2 | 2 |
| 8 | 2 | 2 |
| 9 | 2 | 1 |
| Decision | 0 | 1 |
| | $\mathsf{v_1}$ | $\mathsf{v_2}$ |

Fig. 10. Relation between the genetic algorithm individual encoding and a rule. Matrix A has two rules.

Once the individual is defined, the population which consists in *numpop* individuals stored in a matrix A of dimension $10 \times numpop$ can also be defined. For the vehicle navigation, the value of *numpop* equals to 100 was adopted. Since each column of matrix A represents individuals/solutions, then every element $A(i,j) = k$ means what situation ($i$ equals 1 to 9) or decision ($i$ equals to 10) for the rule $j$ should be made if the solution given from individual $j$ is chosen. For example, $A(1,1) = 2$ means that the individual/solution 1 vehicle radar does not care (value 2) whether there is or not an object in sector 1 of the vehicle radar.

**(B) Fitness evaluation and assignment:**

The fitness evaluation or *fitness* is responsible to rank the individual among the population in a generation t. Then, the fitness is constructed in manner that the solutions with less credit measure have the higher fitness value. This explanation justifies why the fitness for each individual $A_i$ is calculated with Eq. (6), given as follows:

$$Fit(A_i) = f(A_i) \tag{6}$$

Where: $f(A_i)$ is the fitness evaluation as defined by the credit value $C_{it}$, and it corresponds to the number of steps performed with rule $i$ without a collision, according to the rule stored in vector $A_i$, through the environment.

### (C) Individual Selection for the next generation:

The population formation process employed is the *"Roulette Wheel"*, in order words, a random raffle where the probability $Q(A_i)$ to choose the individual $A_i$ for the next generation, in a population with $b$ individuals, is used. The value $Q(A_i)$ can be obtained by using Eq. (7).

$$Q(A_i) = Fit(A_i) / \sum_{i=1}^{numpop} (Fit(A_i)) \qquad (7)$$

The better individual of the actual generation is always kept to be on the next generation.

### (D) Crossover OX:

The Crossover operators try to generate new individuals to form the next population by combining information from past generations that is present in the individuals. This work used two crossover operators which is described as follows.

Two individuals, $A_1$ and $A_2$, with $N$ elements are randomly chosen from a population in generation t. Then, an integer number $\delta$ in the interval [1, $N$-1] is drawn and elements of $A_1$ that are in positions $\delta$ until $N$ are exchanged with the elements of $A_2$ that are in positions $\delta$ until $N$. This exchange will produce two new individuals, $nA_1$ and $nA_2$, that can appear in the next generation.

### (E) Mutation operator:

The mutation operator modifies *10\*numpop\*pm* elements of matrix A, where *pm* is the percentage of total bits to be muted (mutated). The selection of which element $A_{ij}$ to be mutated consist in randomly select the line index and the column index, and then change it. Fig. 11 shows how the genetic algorithm components described before are combined.

One important observation is that the best individual of generation $A_t$ is always kept for the next generation $A_{t+1}$. Furthermore, the size of subpopulations $As_1{}^t$ and $As_2{}^t$ are the same and is equal to 5% of the total population (*numpop*).

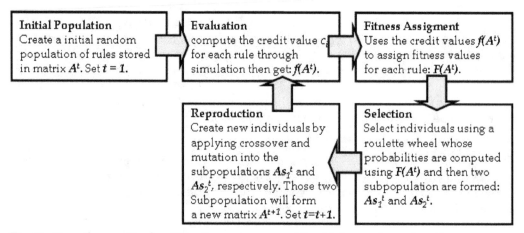

Fig. 11. How the genetic algorithm elements are combined into a unique algorithm.

## 6. Tests and results

Table 1 summarizes the description of the parameters necessary to perform tests with the developed approach described in sections 4 and 5.

| Parameter | Description |
|---|---|
| $C_{i0}$ | Initial credit assigment for a set of rules. |
| $k$ | reinformecent learning rate parameter. |
| $pb0$ | percentage of digits equals to 0 for the initial population. |
| $pb1$ | percentage of digits equals to 1 for the initial population. |
| $pb2$ | percentage of digits equals to 2 for the initial population. |
| $numpop$ | number of individuals in the genetic algorithm population. |
| $pc$ | Crossover rate. |
| $pm$ | Mutation rate. |
| $pg$ | size of the subpopulation in terms of the A matrix (5%). |
| $M$ | number of collisions before applying the genetic algorithm (10). |
| $N$ | number of genetic algorithm aplication(14). |

Table 1. Parameters used for the developed approach and their meanings.

It was also tested two environments whose obstacles initial position are shown in Fig. 12.
As shown in Fig. 12, the obstacles could be clustered in two groups: the obstacles that delimits the environment bounds have their position fixed and the obstacles that are in the interior of the environment, whose positions are actualized in a different manner for each environment:

- Environment 1: The moving obstacles actualize their x-positions according to Eq. (8).

$$x_i^t = x_i^t + \varepsilon \qquad (8)$$

Where: $x_i^t$ is the x-position for obstacle $i$ in step $t$ of the vehicle, $\varepsilon$ is a random uniform variable in the interval [0, 1].

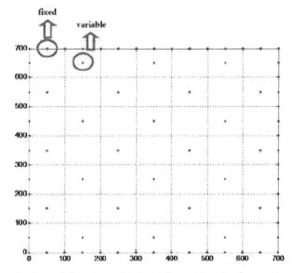

Fig. 12. Detailed description of the obstacles initial position in the environment.

- Environment 2: The user could also specify a positive value to a *radius* parameter and the moving obstacles actualize their x and y-positions according to Eq. (9).

$$x_i^t = x_i^t + radius * \cos(x_i^t)$$
$$y_i^t = y_i^t + radius * \sin(y_i^t)$$

$$(9)$$

For all tests performed in environment 2, the radius value was fixed in 5. Five runs with the values listed in Table 2 had been performed in order to carry the best set of values for all the parametes listed in Table 1. Each row in Table 2 corresponds to a new configuration.

| Set | $C_{i0}$ | k | Pb0 | Pb1 | Pb2 | numpop | pc | pm | pg | M | N | Environment |
|-----|------|---|------|------|-----|--------|----|----|----|----|----|-------------|
| 1 | 20 | 2 | 2.5% | 2.5% | 95% | 100 | 30 | 1 | 5 | 10 | 14 | 1 |
| 2 | 20 | 2 | 2.5% | 2.5% | 95% | 100 | 1 | 30 | 5 | 10 | 14 | 1 |
| 3 | 20 | 2 | 2.5% | 2.5% | 95% | 100 | 30 | 1 | 5 | 20 | 5 | 1 |
| 4 | 20 | 2 | 2.5% | 2.5% | 95% | 100 | 1 | 30 | 5 | 20 | 5 | 1 |
| 5 | 20 | 2 | 10% | 10% | 80% | 100 | 30 | 1 | 5 | 20 | 5 | 1 |
| 6 | 20 | 2 | 10% | 10% | 80% | 100 | 1 | 30 | 5 | 20 | 5 | 1 |
| 7 | 20 | 2 | 2.5% | 2.5% | 95% | 100 | 1 | 30 | 5 | 20 | 5 | 2 |
| 8 | 20 | 2 | 10% | 10% | 80% | 100 | 30 | 1 | 5 | 20 | 5 | 2 |

Table 2. List of values assigned for each parameter.

Each row of Table 3 corresponds to the number of the steps that could be taken with the final set of rules obtained after the application of the scheme described in Fig. 9 using the parameters described in Table 2 and the **reinforcement learning** that follows Eq. (3).

| Set | Test 1 | Test 2 | Test 3 | Test 4 | Test 5 | Best Number |
|-----|--------|--------|--------|--------|--------|-------------|
| 1 | 16 | 7 | 7 | 7 | 7 | 16 |
| 2 | 7 | 110 | 7 | 7 | 80 | 110 |
| 3 | 28 | 57 | 78 | 16 | 7 | 78 |
| 4 | 177 | 32 | 48 | 21 | 18 | 177 |
| 5 | 151 | 18 | 15 | 32 | 116 | 151 |
| 6 | 34 | 32 | 19 | 32 | 106 | 106 |
| 7 | 32 | 163 | 20 | 106 | 84 | 163 |
| 8 | 38 | 15 | 31 | 108 | 31 | 108 |

Table 3. The number of maximum steps without a collision performed by the best set of rules produced by each set of parameters (described in Table 2) and Eq. (3).

In Table 3, the first column indicates the set of parameters and the corresponding obtained set of rules which lead to the number of maximum steps without a collision described in the second column until the sixth column. The results for the set of parameters 1 to 6 show that for environment 1 the best choices of parameters are sets 4 and 5. Those two sets were also tested in environment 2 in order to verify set parameters robustness in order to produce adequate rules for different environments. Configuration 7 had the best performance and three tests were sucessfull in producing rules which could make the vehicle able to take more than 80 steps without a collision.

The new proposed reinforcement learning formula proposed by Eq. (4) and (5) was also tested  for $np = 20$ for the parameter values showed in Table 2.

| Set | Test 1 | Test 2 | Test 3 | Test 4 | Test 5 | Best Number |
|-----|--------|--------|--------|--------|--------|-------------|
| 1 | 20 | 7 | 15 | 7 | 11 | 20 |
| 2 | 7 | 121 | 7 | 7 | 105 | 121 |
| 3 | 36 | 20 | 27 | 7 | 41 | 41 |
| 4 | 37 | 279 | 339 | 296 | 15 | 339 |
| 5 | 15 | 205 | 100 | 126 | 15 | 205 |
| 6 | 84 | 34 | 31 | 617 | 29 | 617 |
| 7 | 32 | 34 | 16 | 37 | 12 | 37 |
| 8 | 31 | 33 | 165 | 31 | 77 | 165 |

Table 4. The number of maximum steps without a collision performed by the best set of rules produced by each set of parameters (described in Table 2) and Eq. (4) and (5).

Fig. 13 shows the simulation of the vehicle control navigation with the best set of rules obtained from the fourth set of parameters.

The data in Table 4 also attest the better performance of the new reinforcement learning scheme proposed by Eq. (4) and (5) (see Table 4) since almost all set of parameters (expections are 3 and 7) created a set of rules with a number of steps without a collision which is bigger than the set of rules created by Eq. (3) (see Table 3).

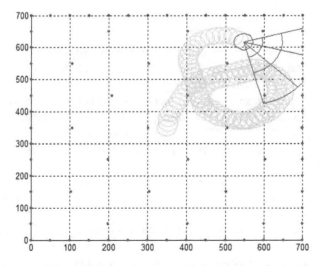

Fig. 13. The simulation of the vehicle navigation with best rules obtained with set 4.

## 7. Conclusions and future works

This Chapter discussed mathematical techniques that could help the ultimate unmanned aerial vehicle (UAVs) objective, which is the development of rules to replace human pilots by altering machines decisions in order to make decisions like humans do. For this purpose, several tools related with artificial intelligence could be employed such as expert systems, neural networks, machine learning and natural language processing (HAYKIN, 2009).

Two methodologies, reinforcement learning and genetic algorithms, are described and combined. The methodologies had been implemented in a Matlab and were successfully applied in order to create rules that keep the vehicle away from the obstacles from random initial rules. This Chapter also described how the ideas developed by (STAFYLOPATIS, 1998) could be modified, particularly the setting of parameters and a new reinforcement learning methodology, to provide a better integration of the Genetic Algorithms (GAs) and Reinforcement Learning (RL) which produce rules for automatic guided vehicles with a better performance.

The main contributions of this Chapter were: the vehicle, its sensors, and also the environment for training which are different from the ones presented in (STAFYLOPATIS, 1998); a new equation for the reinforcement learning was proposed; the influence of the parameters that control the production of automatic generation of rules for vehicle control navigation are also tested. The results presented attested the better performance of the new reinforcement learning scheme proposed, implemented in Matlab. For future works, it could be applied a genetic algorithm for finding the better configuration of the parameters for the combined scheme of reinforcement learning and the genetic algorithm used to find the rules.

For future works more tests could be performed by using other equations for the reinforcement learning process and also other methods like Beam Search (SABUNCUOGLU and BAVIZ, 1999; DELLA CROCE and T'KINDT, 2002) could replace the Genetic Algorithm. A more detailed and complex decisions could also be incorporated such as the possibility of increasing or reducing the velocity of the vehicle.

## 8. References

Della Croce, F.; T'Kind, V., A Recovering Beam Search Algorithm for the One-Machine Dynamic Total Completation Time Scheduling Problem, Journal of the Operational Research Society, vol. 54, p. 1275-1280, 2002.

Haykin, , S. Neural Networks and Learning Machines, 3rd edition, Prentice Hall, 2009.

Sutton, R.S., Barto, A.G., *Reinforcement Learning: An Introduction*, MIT Press, 1998.

Stafylopatis, A., Blekas, K., Autonomous vehicle navigation using evolutionary reinforcement learning, European Journal of Operational Research, Vol. 108(2), p. 306-318, 1998.

Shim, D. H, Kim, H. J., Sastry, S., Hierarchical Control System Synthesis for Rotorcraft-based Unmanned Aerial Vehicles, AIAA Guidance, Navigation, and Control Conference and Exhibit, v. 1, p. 1-9, 2000.

Holland, J. H., Adaptation in natural and artificial systems. The University of Michigan Press, 1975.

Michalewicz, Z. Genetic Algorithms + Data Structures = Evolution Programs, 3rd edition, Springer-Verlag, 1996.

Sabuncuoglu, I., Baviz, M., Job Shop Scheduling with Beam Search, European Journal of Operational Research, vol. 118, pp. 390-412.

# 9

# Implementation of a New Quasi-Optimal Controller Tuning Algorithm for Time-Delay Systems

Libor Pekař and Roman Prokop
*Tomas Bata University in Zlín*
*Czech Republic*

## 1. Introduction

Systems and models with dead time or aftereffect, also called hereditary, anisochronic or time-delay systems (TDS), belonging to the class of infinite dimensional systems have been largely studied during last decades due to their interesting and important theoretical and practical features. A wide spectrum of systems in natural sciences, economics, pure informatics etc., both real-life and theoretical, is affected by delays which can have various forms; to name just a few the reader is referred e.g. to (Górecki et al., 1989; Marshall et al., 1992; Kolmanovskii & Myshkis, 1999; Richard, 2003; Michiels & Niculescu, 2008; Pekař et al., 2009) and references herein. Linear time-invariant dynamic systems with distributed or lumped delays (LTI-TDS) in a single-input single-output (SISO) case can be represented by a set of functional differential equations (Hale & Verduyn Lunel, 1993) or by the Laplace transfer function as a ratio of so-called quasipolynomials (El'sgol'ts & Norkin, 1973) in one complex variable $s$, rather than polynomials which are usual in system and control theory. Quasipolynomials are formed as linear combinations of products of $s$-powers and exponential terms. Hence, the Laplace transform of LTI-TDS is no longer rational and so-called meromorphic functions have to be introduced. A significant feature of LTI-TDS is (in contrast to undelayed systems ) its infinite spectrum and transfer function poles decide - except some cases of distributed delays, see e.g. (Loiseau, 2000) - about the asymptotic stability as in the case of polynomials.

It is a well-known fact that delay can significantly deteriorate the quality of feedback control performance, namely stability and periodicity. Therefore, design a suitable control law for such systems is a challenging task solved by various techniques and approaches; a plentiful enumeration of them can be found e.g. in (Richard, 2003). Every controller design naturally requires and presumes a controlled plant model in an appropriate form. A huge set of approaches uses the Laplace transfer function; however, it is inconvenient to utilize a ratio of quasipolynomials especially while natural requirements of internal (impulse-free modes) and asymptotic stability of the feedback loop and the feasibility and causality of the controller are to be fulfilled.

The meromorphic description can be extended to the fractional description, to satisfy requirements above, so that quasipolynomials are factorized into proper and stable meromorphic functions. The ring of stable and proper quasipolynomial (RQ) meromorphic functions ($\mathbf{R}_{MS}$) is hence introduced (Zítek & Kučera, 2003; Pekař & Prokop, 2010). Although the ring can be used for a description of even neutral systems (Hale & Verduyn Lunel, 1993), only systems with so-called retarded structure are considered as the admissible class of systems in this contribution. In contrast to many other algebraic approaches, the ring enables to handle systems with non-commensurate delays, i.e. it is not necessary that all system delays can be expressed as integer multiples of the smallest one. Algebraic control philosophy in this ring then exploits the Bézout identity, to obtain stable and proper controllers, along with the Youla-Kučera parameterization for reference tracking and disturbance rejection.

The closed-loop stability is given, as for delayless systems, by the solutions of the characteristic equation which contains a quasipolynomial instead of a polynomial. These infinite many solutions represent closed-loop system poles deciding about the control system stability. Since a controller can have a finite number of coefficients representing selectable parameters, these have to be set to distribute the infinite spectrum so that the closed-loop system is stable and that other control requirements are satisfied.

The aim of this chapter is to describe, demonstrate and implement a new quasi-optimal pole placement algorithm for SISO LTI-TDS based on the quasi-continuous pole shifting – the main idea of which was presented in (Michiels et al., 2002) - to the prescribed positions. The desired positions are obtained by overshoot analysis of the step response for a dominant pair of complex conjugate poles. A controller structure is initially calculated by algebraic controller design in $\mathbf{R}_{MS}$. Note that the maximum number of prescribed poles (including their multiplicities) equals the number of unknown parameters. If the prescribed roots locations can not be reached, the optimizing of an objective function involving the distance of shifting poles to the prescribed ones and the roots dominancy is utilized. The optimization is made via Self-Organizing Migration Algorithm (SOMA), see e.g. (Zelinka, 2004). Matlab m-file environment is utilized for the algorithm implementation and, consequently, results are tested in Simulink on an attractive example of unstable SISO LTI-TDS.

The chapter is organized as follows. In Section 2 a brief general description of LTI-TDS is introduced together with the coprime factorization for the $\mathbf{R}_{MS}$ ring representation. Basic ideas of algebraic controller design in $\mathbf{R}_{MS}$ with a simple control feedback are presented in Section 3. The main and original part of the chapter – pole-placement shifting based tuning algorithm – is described in Section 4. Section 5 focuses SOMA and its utilization when solving the tuning problem. An illustrative benchmark example is presented in Section 6.

## 2. Description of LTI-TDS

The aim of this section is to present possible models of LTI-TDS; first, that in time domain using functional differential equations, second, the transfer function (matrix) via the Laplace transform. Then, the latter concept is extended so that an algebraic description in a special ring is introduced. Note that for the further purpose of this chapter the state-space functional description is useless.

### 2.1 State-space model

A LTI-TDS system with both input-output and internal (state) delays, which can have point (lumped) or distributed form, can be expressed by a set of functional differential equations

$$\frac{dx(t)}{dt} = \sum_{i=1}^{N_H} \mathbf{H}_i \frac{dx(t-\eta_i)}{dt} + \mathbf{A}_0 \mathbf{x}(t) + \sum_{i=1}^{N_A} \mathbf{A}_i \mathbf{x}(t-\eta_i) + \mathbf{B}_0 \mathbf{u}(t)$$

$$+ \sum_{i=1}^{N_B} \mathbf{B}_i \mathbf{u}(t-\eta_i) + \int_0^L \mathbf{A}(\tau) \mathbf{x}(t-\tau) d\tau + \int_0^L \mathbf{B}(\tau) \mathbf{u}(t-\tau) d\tau \qquad (1)$$

$$\mathbf{y}(t) = \mathbf{C}\mathbf{x}(t)$$

where $\mathbf{x} \in \mathbb{R}^n$ is a vector of state variables, $\mathbf{u} \in \mathbb{R}^m$ stands for a vector of inputs, $\mathbf{y} \in \mathbb{R}^l$ represents a vector of outputs, $\mathbf{A}_i$, $\mathbf{A}(\tau)$, $\mathbf{B}_i$, $\mathbf{B}(\tau)$, $\mathbf{C}$, $\mathbf{H}_i$ are matrices of appropriate dimensions, $0 \leq \eta_i \leq L$ are lumped (point) delays and convolution integrals express distributed delays (Hale & Verduyn Lunel, 1993; Richard, 2003; Vyhlídal, 2003). If $\mathbf{H}_i \neq \mathbf{0}$ for any $i = 1,2,...N_H$, model (1) is called neutral; on the other hand, if $\mathbf{H}_i = \mathbf{0}$ for every $i = 1,2,...N_H$, so-called retarded model is obtained. It should be noted that the state of model (1) is given not only by a vector of state variables in the current time instant, but also by a segment of the last model history (in functional Banach space) of state and input variables

$$\mathbf{x}(t+\tau), \ \mathbf{u}(t+\tau), \ \tau \in \langle -L, 0 \rangle \qquad (2)$$

Convolution integrals in (1) can be numerically approximate by summations for digital implementation; however, this can destabilize even a stable system. Alternatively, one can integrate (1) and add a new state variable to obtain derivations on the right-hand side only. In the contrary, the model can also be expressed in more consistent functional form using Riemann-Stieltjes integrals so that both lumped and distributed delays are under one convolution. For further details and other state-space TDS models the reader is referred to (Richard, 2003).

## 2.2 Input-output model
This contribution is concerned with retarded delayed systems in the input-output formulation governed by the Laplace transfer function matrix (considering zero initial conditions) as in (3). Hence, in the SISO case (we are concerning about here), the transfer function is no longer rational, as for conventional delayless systems, and a meromorphic function as a ratio of retarded quasipolynomials (RQ) is obtained instead.

$$\mathbf{Y}(s) = \mathbf{G}(s)\mathbf{U}(s) = \frac{\Gamma_1 \Gamma_2}{\Delta} \mathbf{U}(s)$$

$$\Delta = \det\left[ s\mathbf{I} - \mathbf{A}_0 - \sum_{i=1}^{N_A+1} \mathbf{A}_i \exp(-s\eta_i) - \int_0^L \mathbf{A}(\tau)\exp(-s\tau)d\tau \right]$$

$$\Gamma_1 = \mathbf{C}\,\mathrm{adj}\left[ s\mathbf{I} - \mathbf{A}_0 - \sum_{i=1}^{N_A+1} \mathbf{A}_i \exp(-s\eta_i) - \int_0^L \mathbf{A}(\tau)\exp(-s\tau)d\tau \right] \qquad (3)$$

$$\Gamma_2 = \mathbf{B}_0 + \sum_{i=1}^{N_B+1} \mathbf{B}_i \exp(-s\eta_i) + \int_0^L \mathbf{B}(\tau)\exp(-s\tau)d\tau$$

A (retarded or neutral) quasipolynomial of degree $n$ has the generic form

$$q(s) = s^n + \sum_{i=0}^{n}\sum_{j=1}^{h} x_{ij}s^i \exp\left(-\vartheta_{ij}s\right), \vartheta_{ij} \geq 0 \tag{4}$$

where $x_{nj} \neq 0$ in the neutral case for some $j$, whereas a RQ owns $x_{nj} = 0$ for all $j$.

However, the transfer function representation in the form of a ratio of two quasipolynomials is not suitable in order to satisfy controller feasibility, causality and closed-loop (Hurwitz) stability (Loiseau 2000; Zítek & Kučera, 2003). Rather more general approaches utilize a field of fractions where a transfer function is expressed as a ratio of two coprime elements of a suitable ring. A ring is a set closed for addition and multiplication, with a unit element for addition and multiplication and an inverse element for addition. This implies that division is not generally allowed.

### 2.3 Plant description in $R_{MS}$ ring

A powerful algebraic tool ensuring requirements above is a ring of stable and proper RQ-meromorphic functions ($R_{MS}$). Since the original definition of $R_{MS}$ in (Zítek & Kučera, 2003) does not constitute a ring, some minor changes in the definition was made in (Pekař & Prokop, 2009). Namely, although the retarded structure of TDS is considered only, the minimal ring conditions require the use of neutral quasipolynomials at least in the numerator as well.

An element $T(s) \in R_{MS}$ is represented by a proper ratio of two quasipolynomials

$$T(s) = \frac{y(s)}{x(s)} \tag{5}$$

where a denominator $x(s)$ is a quasipolynomial of degree $n$ and a numerator can be factorized as

$$y(s) = \tilde{y}(s)\exp(-\tau s) \tag{6}$$

where $\tilde{y}(s)$ is a quasipolynomial of degree $l$ and $\tau \geq 0$. $T(s)$ is stable, which means that there is no pole $s_0$ such that $\mathrm{Re}\{s_0\} \geq 0$; in other words, all roots of $x(s)$ with $\mathrm{Re}\{s_0\} \geq 0$ are those of $y(s)$. Moreover, the ratio is proper, i.e. $l \leq n$.

Thus, $T(s)$ is analytic and bounded in the open right half-plane, i.e.

$$\sup_{\mathrm{Re}\{s\} \geq 0} |T(s)| < \infty \tag{7}$$

As mentioned above, in this contribution only retarded systems are considered, i.e. $x(s)$, $y(s)$ are RQs. Let the plant be initially described as

$$G(s) = \frac{b(s)}{a(s)} \tag{8}$$

where $a(s)$, $b(s)$ are RQs. Hence, using a coprime factorization, a plant model has the form

$$G(s) = \frac{B(s)}{A(s)} \tag{10}$$

where $A(s), B(s) \in \mathbf{R}_{MS}$ are coprime, i.e. there does not exist a non-trivial (non-unit) common factor of both elements. Note that a system of neutral type can induce problem since there can exist a coprime pair $A(s), B(s)$ which is not, however, Bézout coprime – which implies that the system can not be stabilized by any feedback controller admitting the Laplace transform, see details in (Loiseau et al., 2002).

## 3. Controller design in RMS

This section outlines controller design based on the algebraic approach in the $\mathbf{R}_{MS}$ ring satisfying the inner Hurwitz (Bounded Input Bounded Output - BIBO) stability of the closed loop, controller feasibility, reference tracking and disturbance rejection.

For algebraic controller design in $\mathbf{R}_{MS}$ it is initially supposed that not only the plant is expressed by the transfer function over $\mathbf{R}_{MS}$ but a controller and all system signals are over the ring. As a control system, the common negative feedback loop as in Fig. 1 is chosen for the simplicity, where $W(s)$ is the Laplace transform of the reference signal, $D(s)$ stands for that of the load disturbance, $E(s)$ is transformed control error, $U_0(s)$ expresses the controller output (control action), $U(s)$ represents the plant input, and $Y(s)$ is the plant output controlled signal in the Laplace transform. The plant transfer function is depicted as $G(s)$, and $G_R(s)$ stands for a controller in the scheme.

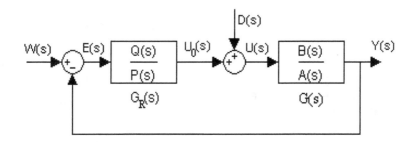

Fig. 1. Simple control feedback loop

Control system external inputs have forms

$$W(s) = \frac{H_W(s)}{F_W(s)}, \ D(s) = \frac{H_D(s)}{F_D(s)} \tag{11}$$

where $H_W(s)$, $H_D(s)$, $F_W(s)$, $F_D(s) \in \mathbf{R}_{MS}$.

The following basic transfer functions can be derived in the control system in general

$$G_{WY}(s) = \frac{Y(s)}{W(s)} = \frac{B(s)Q(s)}{M(s)}, \ G_{DY}(s) = \frac{Y(s)}{D(s)} = \frac{B(s)P(s)}{M(s)}$$

$$G_{WE}(s) = \frac{E(s)}{W(s)} = \frac{A(s)P(s)}{M(s)}, \ G_{DE}(s) = \frac{E(s)}{D(s)} = -\frac{B(s)P(s)}{M(s)} \tag{12}$$

where

$$G_R(s) = \frac{Q(s)}{P(s)} \tag{13}$$

$$M(s) = A(s)P(s) + B(s)Q(s) \tag{14}$$

and $Q(s)$, $P(s)$ are from $\mathbf{R}_{MS}$ and the fraction (13) is (Bézout) coprime (or relatively prime). The numerator of $M(s) \in \mathbf{R}_{MS}$ agrees to the characteristic quasipolynomial of the closed loop.
Following subsections describes briefly how to provide the basic control requirements.

### 3.1 Stabilization
According to e.g. (Kučera, 1993; Zítek & Kučera, 2003), the closed-loop system is stable if and only if there exists a pair $P(s), Q(s) \in \mathbf{R}_{MS}$ satisfying the Bézout identity

$$A(s)P(s) + B(s)Q(s) = 1 \tag{15}$$

a particular stabilizing solution of which, $P_0(s), Q_0(s)$, can be then parameterized as

$$\begin{aligned} P(s) &= P_0(s) \pm B(s)T(s) \\ Q(s) &= Q_0(s) \mp A(s)T(s), \ T(s) \in \mathbf{R}_{MS} \end{aligned} \tag{16}$$

Parameterization (16) is used to satisfy remaining control and performance requirements.

### 3.2 Reference tracking and disturbance rejection
The question is how to select $T(s) \in \mathbf{R}_{MS}$ in (16) so that tasks of reference tracking and disturbance rejection are accomplished. The key lies in the form of $G_{WE}(s)$ and $G_{DY}(s)$ in (12). Consider the limits

$$\begin{aligned} \lim_{t \to \infty} y_D(t) &= \lim_{s \to 0} sY_D(s) = \lim_{s \to 0} sG_{DY}(s)D(s) \\ &= \lim_{s \to 0} sB(s)P(s)\frac{H_D(s)}{F_D(s)} \end{aligned} \tag{17}$$

$$\begin{aligned} \lim_{t \to \infty} e_W(t) &= \lim_{s \to 0} sE_W(s) = \lim_{s \to 0} sG_{WE}(s)W(s) \\ &= \lim_{s \to 0} sA(s)P(s)\frac{H_W(s)}{F_W(s)} \end{aligned} \tag{18}$$

where $\cdot_D$ means that the output is influenced only by the disturbance, and symbol $\cdot_W$ expresses that the signal is a response to the reference. Limit (17) is zero if $\lim_{s \to 0} Y_D(s) < \infty$ and $Y_D(s)$ is analytic in the open right half-plane. Moreover, for the feasibility of $y_D(t)$, $Y_D(s)$ must be proper. This implies that the disturbance is asymptotically rejected if $Y_D(s) \in \mathbf{R}_{MS}$. Similarly, the reference is tracked if $E_W(s) \in \mathbf{R}_{MS}$.
In other words, $F_D(s)$ must divide the product $B(s)P(s)$ in $\mathbf{R}_{MS}$, and $A(s)P(s)$ must be divisible by $F_W(s)$ in $\mathbf{R}_{MS}$. Details about divisibility in $\mathbf{R}_{MS}$ can be found e.g. in (Pekař & Prokop, 2009). Thus, if neither $B(s)$ has any common unstable zero with $F_D(s)$ nor $A(s)$

has any common unstable zero with $F_W(s)$, one has to set all unstable zeros of $F_D(s)$ and $F_W(s)$ (with corresponding multiplicities) as zeros of $P(s)$. Note that zeros mean zero points of a whole term in $\mathbf{R}_{MS}$, not only of a quasipolynomial numerator. Unstable zeros agrees with those with $\mathrm{Re}\{s\} \geq 0$.

## 4. Pole-placement shifting based controller tuning algorithm

In this crucial section, the idea of a new pole-placement shifting based controller tuning algorithm (PPSA) is presented. Although some steps of PPSA are taken over some existing pole-shifting algorithms, the idea of connection with pole placement and the SOMA optimization is original.

### 4.1 Overview of PSSA

We first give an overview of all steps of PPSA and, consequently, describe each in more details. The procedure starts with controller design in $\mathbf{R}_{MS}$ introduced in the previous section. The next steps are as follows:

1.  Calculate the closed-loop reference-to-output transfer function $G_{WY}(s)$. Let $l_{num}$ and $l_{den}$, respectively, be numbers of unknown (free, selectable) real parameters of the numerator and denominator, respectively. Sign $l = l_{num} + l_{den}$.
2.  Choose a simple model of a stable LTI system in the form of the transfer function $G_{WY,m}(s)$ with a numerator of degree $n_{num}$ and the denominator of degree $n_{den}$. Calculate step response maximum overshoots of the model for a suitable range of its $n_{num}$ zeros and $n_{den}$ poles (including their multiplicities). If $\bar{n}_{num} \leq n_{num}$ and $\bar{n}_{den} \leq n_{den}$, respectively, are numbers of all real zeros (poles) and pairs of complex conjugate zeros (poles) of the model, it must hold that $\bar{n}_{num} \leq l_{num}$ and $\bar{n}_{den} \leq l_{den}$, respectively.
3.  Prescribe all poles and zeros of the model with respect to calculated maximum overshoots (and maximal overshoot times). If the poles and zeros are dominant (i.e. the rightmost), the procedure is finished. Otherwise do following steps.
4.  Shift the rightmost (or the nearest) zeros and poles to the prescribed locations successively. If the number of currently shifted poles and conjugate pairs $\bar{n}_{den} \leq n_{sp} \leq l_{den}$ is higher then $\bar{n}_{den}$, try to move the rest of dominant (rightmost) poles to the left. The same rule holds for shifted zeros, analogously.
5.  If all prescribed poles and zeros are dominant, the procedure is finished. Otherwise, select a suitable cost function reflecting the distance of dominant poles (zeros) from prescribed positions and distances of spectral abscissas of both, prescribed and dominant poles (zeros).
6.  Minimize the cost function, e.g. via SOMA.

Now look at these steps of the algorithm at great length.

### 4.2 Characteristic quasipolynomial and characteristic entire function

Algebraic controller design in the $\mathbf{R}_{MS}$ ring introduced in Section 3 results in a controller owning the transfer function $G_R(s)$ containing a finite number of unknown (free, selectable) parameters. The task of PPSA is to set these parameters so that the possibly infinite spectrum of the closed loop has dominant (rightmost) poles located in (or near by) the prescribed positions. If possibly, one can prescribe and place dominant zeros as well. Note that controller design in $\mathbf{R}_{MS}$ using the feedback system as in Fig. 1 results in infinite spectrum of the feedback if the controlled plant is unstable.

If the (quasi)polynomial numerator and denominator of $G(s)$ have no common roots in the open right-half plane, the closed-loop spectrum is given entirely by roots of the numerator $m(s)$ of $M(s)$, the so called characteristic quasipolynomial. In the case of distributed delays, $G(s)$ has some common roots with $\mathrm{Re}\{s\} \geq 0$ in both, the numerator and denominator, and these roots do not affect the system dynamics since they cancel each other. In this case, the spectrum is given by zeros of the entire function $m(s)/m_U(s)$, i.e. the characteristic entire function, where $m_U(s)$ is a (quasi)polynomial the only roots of which are the common unstable roots.

The (quasi)polynomial denominator of $G_{WY}(s)$ agrees with $m(s)$. Its role is much more important than the role of the numerator of $G_{WY}(s)$ since the closed-loop zeros does not influence the stability. In the light of this fact, the setting of closed-loop poles has the priority. Therefore, one has to set $l_{den}$ free denominator parameters first. Free (selectable) parameters in the numerator of $G_{WY}(s)$ are to be set only if there exist those which are not contained in the denominator. The number of such "additional" parameters is $l_{num}$.

### 4.3 Closed-loop model and step response overshoots

The task now is how to prescribe the closed-loop poles appropriately. We choose a simple finite-dimensional model of the reference-to-output transfer function and find its maximum overshoots and overshoot times for a suitable range of the model poles.

Let the prescribed (desired) closed-loop model be of the transfer function

$$G_{WY,m}(s) = k_1 \frac{b_1 s + b_0}{s^2 + a_1 s + a_0} = k_2 \frac{s - z_1}{(s - s_1)(s - \bar{s}_1)} \tag{19}$$

where $k_1, k_2, b_1, b_0, a_1, a_0 \in \mathbb{R}$ are model parameters $z_1 \in \mathbb{R}^-$ stands for a model zero and $s_1 \in \mathbb{C}^-$ is a model stable pole where $\bar{s}_1$ expresses its complex conjugate. To obtain the unit static gain of $G_{WY,m}(s)$ it must hold true

$$k_1 = \frac{a_0}{b_0}, k_2 = -\frac{|s_1|^2}{z_1} \tag{20}$$

Sign $s_1 = \alpha + \omega\mathrm{j}, \alpha < 0, \omega \geq 0$ and calculate the impulse function $g_{WY,m}(t)$ of $G_{WY,m}(s)$ using the Matlab function *ilaplace* as

$$g_{WY,m}(t) = k_2 \exp(\alpha t)\left[\cos(\omega t) - \frac{z_1 - \alpha}{\omega}\sin(\omega t)\right] \tag{21}$$

Since $i_{WY,m}(t) = h'_{WY,m}(t)$, where $h_{WY,m}(t)$ is the step response function, the necessary condition for the existence of a step response overshoot at time $t_O$ is

$$i_{WY,m}(t_O) = 0, t_O > 0 \tag{22}$$

The condition (22) yields these two solutions: either $t_O \to -\infty$ (which is trivial) or

$$t_O = \frac{1}{\omega}\arccos\left(\pm\frac{|\alpha - z_1|}{\sqrt{(\alpha - z_1)^2 + \omega^2}}\right) \qquad (23)$$

when considering $\arccos(x) \in [0, \pi]$. Obviously, (23) has infinitely many solutions. If $\alpha < 0, z_1 < 0$, the maximum overshoot occurs at time

$$t_{\max} = \min(t_O) \qquad (24)$$

One can further calculate the step response function $h_{WY,m}(t)$ as

$$h_{WY,m}(t) = \frac{k_2}{|s_1|^2}\left[\exp(\alpha t)\left(z_1 \cos(\omega t) - \frac{z_1\alpha - |s_1|^2}{\omega}\sin(\omega t)\right) - z_1\right] \qquad (25)$$

Define now the maximum relative overshoot as

$$\Delta h_{WY,m,\max} := \frac{h_{WY,m}(t_{\max}) - h_{WY,m}(\infty)}{h_{WY,m}(\infty)} \qquad (26)$$

see Fig. 2.

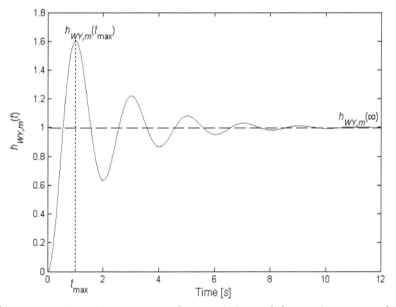

Fig. 2. Reference-to-output step response characteristics and the maximum overshoot

Using definition (26) one can obtain

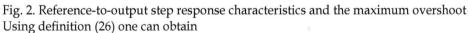

$$\Delta h_{WY,m,\max} = \exp(\alpha t)\left(\frac{-z_1\omega\cos(\omega t) + \left(z_1\alpha - |s_1|^2\right)\sin(\omega t)}{z_1\omega}\right)\Bigg|_{t=t_{\max}} \qquad (27)$$

Obviously, $\Delta h_{WY,m,max}$ is a function of three parameters, i.e. $n_1, \alpha, \omega$, which is not suitable for a general formulation of the maximal overshoot. Hence, let us introduce new parameters $\xi_\alpha, \xi_z$ as

$$\xi_\alpha = -\frac{\alpha}{\omega}, \ \xi_z = -\frac{z_1}{\omega} \tag{28}$$

which give rise from (23), (24) and (27) to

$$\Delta h_{WY,m,max} = \frac{1}{\xi_z} \exp\left(-\xi_\alpha t_{max,norm}\right)\left(-\xi_z \cos\left(t_{max,norm}\right) + \left(\xi_\alpha^2 + 1 - \xi_\alpha \xi_z\right)\sin\left(t_{max,norm}\right)\right)$$

$$t_{max,norm} = \omega t_{max} = \min\left(\arccos\left(\pm \frac{|\xi_\alpha - \xi_\omega|}{\sqrt{\left(\xi_\alpha - \xi_\omega\right)^2 + 1}}\right)\right) \tag{29}$$

where $t_{max,norm}$ represents the normalized maximal overshoot time.

We can successfully use Matlab to display function $\Delta h_{WY,m,max}\left(\xi_\alpha, \xi_z\right)$ and $t_{max,norm}\left(\xi_\alpha, \xi_z\right)$ graphically, for suitable ranges of $\xi_\alpha, \xi_z$ as can be seen from Fig. 3 – Fig. 7.

Recall that model (19) gives rise to $n_{num} = 1, n_{den} = 2, \bar{n}_{num} = 1, \bar{n}_{den} = 1$.

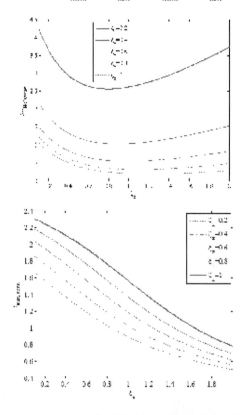

Fig. 3. Maximum overshoots $\Delta h_{WY,m,max}\left(\xi_\alpha, \xi_z\right)$ (a) and normalized maximal overshoot times $t_{max,norm}\left(\xi_\alpha, \xi_z\right)$ (b) for $\xi_\alpha = [0.1, 2]$, $\xi_z = \{0.2, 0.4, 0.6, 0.8, 1\}$.

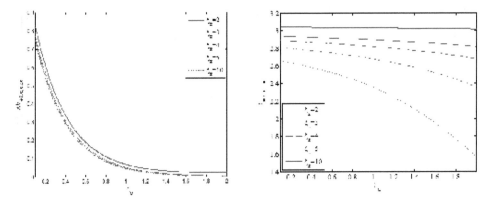

Fig. 4. Maximum overshoots $\Delta h_{WY,m,\max}(\xi_\alpha,\xi_z)$ (a) and normalized maximal overshoot times $t_{\max,norm}(\xi_\alpha,\xi_z)$ (b) for $\xi_\alpha=[0.1,2]$, $\xi_z=\{2,3,4,5,10\}$.

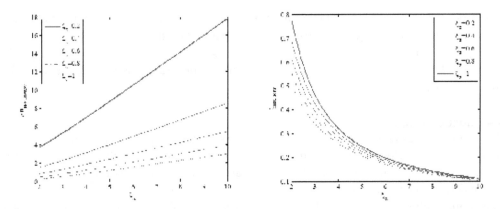

Fig. 5. Maximum overshoots $\Delta h_{WY,m,\max}(\xi_\alpha,\xi_z)$ (a) and normalized maximal overshoot times $t_{\max,norm}(\xi_\alpha,\xi_z)$ (b) for $\xi_\alpha=[2,10]$, $\xi_z=\{0.2,0.4,0.6,0.8,1\}$.

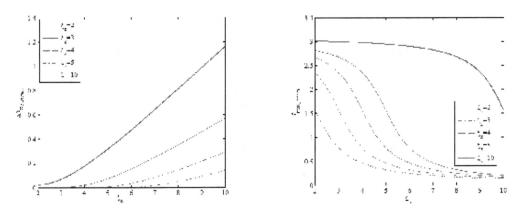

Fig. 6. Maximum overshoots $\Delta h_{WY,m,\max}(\xi_\alpha,\xi_z)$ (a) and normalized maximal overshoot times $t_{\max,norm}(\xi_\alpha,\xi_z)$ (b) for $\xi_\alpha=[2,10]$, $\xi_z=\{2,3,4,5,10\}$.

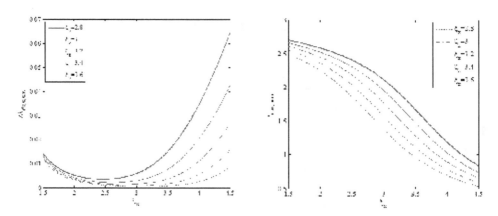

Fig. 7. Maximum overshoots $\Delta h_{WY,m,\max}(\xi_\alpha,\xi_z)$ (a) and normalized maximal overshoot times $t_{\max,norm}(\xi_\alpha,\xi_z)$ (b) for $\xi_\alpha=[1,5,4.5]$, $\xi_z=\{2.8,3,3.2,3.4,3.6\}$ - A detailed view on "small" overshoots.

The procedure of searching suitable prescribed poles can be done e.g. as in the following way. A user requires $\Delta h_{WY,m,\max}=0.03$ (i.e. the maximal overshoot equals 3 %), $\xi_\alpha=4$ (i.e. "the quarter dumping") and $t_{\max}=5\,\text{s}$. Fig. 7 gives approximately $\xi_z=2.9$ which yields $t_{\max,norm}\approx1.2$. These two values together with (28) and (29) result in $s_1=-0.96+0.24\text{j}$, $z_1=-0.7$.

### 4.4 Direct pole placement
This subsection extends step 3 of PPSA from Subsection 4.1. The goal is to prescribe poles and zeros of the closed-loop "at once". The drawback here is that the prescribed poles (zeros) might not be dominant (i.e. the rightmost). The procedure was utilized to LTI-TDS e.g. in (Zítek & Hlava, 2001).

Given quasipolynomial $m(s)$ with a vector $\mathbf{v}=[v_1,v_2,...,v_l]^T\in\mathbb{R}^l$ of $l$ free parameters, the assignment of $n$ prescribed single roots $\sigma_i$, $i=1...n$, can be done via the solution of the set of algebraic equations in the form

$$[m(\sigma_i,\mathbf{v})]_{s=\sigma_i}=0,\,i=1...n \tag{30}$$

In the case of complex conjugate poles, one has to take the real and imaginary part separately as

$$\text{Re}\left\{[m(\sigma_i,\mathbf{v})]_{s=\sigma_i}\right\}=0,\text{Im}\left\{[m(\sigma_i,\mathbf{v})]_{s=\sigma_i}\right\}=0 \tag{31}$$

for every pair of roots.

If a root $\sigma_i$ has the multiplicity $p$, it must be calculated

$$\left[\frac{d^j}{ds^j}m(\sigma_i,\mathbf{v})\right]_{s=\sigma_i}=0,\,j=0...p-1 \tag{32}$$

or

$$\text{Re}\left[\frac{d^j}{ds^j}m(\sigma_i,\mathbf{v})\right]_{s=\sigma_i}=0,\ \text{Im}\left[\frac{d^j}{ds^j}m(\sigma_i,\mathbf{v})\right]_{s=\sigma_i}=0, j=0...p-1 \tag{33}$$

Note that if $m(s)$ is nonlinear with respect to $\mathbf{v}$, one can solve a set on non-linear algebraic equations directly, or to use an expansion

$$m(\sigma_i,\mathbf{v}) \approx m(\sigma_i,\mathbf{v}_0) + \sum_{j=1}^{k}\Delta v_j\left[\frac{\partial m(s,\mathbf{v}_0)}{\partial v_j}\right]_{\substack{s=\sigma_i\\\mathbf{v}=\mathbf{v}_0}} \tag{34}$$

where $\mathbf{v}_0$ means a point in which the expansion is made or an initial estimation of the solution and $\Delta\mathbf{v}=[\Delta v_1,\Delta v_2,...,\Delta v_l]^T$ is a vector of parameters increments. Equations (34) should be solved iteratively, e.g. via the well-known Newton method. Note, furthermore, that the algebraic controller design in $\mathbf{R}_{MS}$ for LTI-TDS results in the linear set (30)-(34) with respect to selectable parameters – both, in the numerator and denominator of $G_{WY}(s)$.

It is clear that a unique solution is obtained only if the set of $n=l$ independent equations is given. If $n<l$, equations (30)-(34) can be solved using the Moore-Penrose (pseudo)inverse

minimizing the norm $\|\mathbf{v}\|_2 = \sum_{i=1}^{k}v_i^2$ , see (Ben Israel & Greville, 1966). Contrariwise, whenever

$n>l$ , it is not possible to place roots exactly and the pseudoinverse provides the minimization of squares of the left-hand sides of (30)-(34).

The methodology described in this subsection is utilized on both, the numerator and denominator.

### 4.5 Continuous poles (zeros) shifting

Once the poles (zeros) are prescribed, it ought to be checked whether these roots are the rightmost. If yes, the PPSA algorithm stops; if not, one may try to shift poles so that the prescribed ones become dominant. There are two possibilities. First, the dominant roots move to the prescribed ones; second, roots nearest to the prescribed ones are shifted – while the rest of the spectrum (or zeros) is simultaneously pushed to the left. The following describes it in more details.

We describe the procedure for the closed-loop denominator and its roots (poles); the numerator is served analogously for all its free parameters which are not included in the denominator. Recall that $l_{den}$ is the number of unknown (selectable) parameters, $n_{den}$ stands for the number of model (prescribed) poles (including their multiplicities), $\bar{n}_{den}$ represents the number of real poles and conjugate pairs of prescribed poles and $n_{sp}$ is the number of currently shifted real poles and conjugate pairs. Generally, it holds that

$$\bar{n}_{den} \le n_{sp} \le l_{den} \tag{35}$$

The idea of continuous poles shifting described below was introduced in (Michiels et al., 2002). Similar procedure which, however, enables to shift less number of poles since $n_{sp} \le l_{den}$ includes every single complex pole instead of a conjugate pair, was investigated in (Vyhlídal, 2003). Roughly speaking, the latter is based on solution of (30) - (34) where $\mathbf{v}_0$ represents the vector of actual controller parameters, $\mathbf{v}=\mathbf{v}_0+\Delta\mathbf{v}$ are new controller parameters and $\sigma_i$ means prescribed poles (in the vicinity of the actual ones) here. Now look at the former methodology in more details.

The approach (Michiels et al., 2002) is based on the extrapolation

$$m(\sigma_i, \mathbf{v}) \approx \underbrace{m(\sigma_i, \mathbf{v}_0)}_{=0} + \Delta\sigma_i \left[ \frac{\partial m(s, \mathbf{v})}{\partial s} \right]_{\substack{s=\sigma_i \\ \mathbf{v}=\mathbf{v}_0}} + \Delta v_j \left[ \frac{\partial m(s, \mathbf{v})}{\partial v_j} \right]_{\substack{s=\sigma_i \\ \mathbf{v}=\mathbf{v}_0}} = 0, \ i = 1...n_{sp}, \ j = 1...l_{den} \quad (36)$$

yielding

$$\frac{\Delta\sigma_i}{\Delta v_j} \approx -\left[ \frac{\partial m(s, \mathbf{v})}{\partial s} \right]_{\substack{s=\sigma_i \\ \mathbf{v}=\mathbf{v}_0}}^{-1} \left[ \frac{\partial m(s, \mathbf{v})}{\partial v_j} \right]_{\substack{s=\sigma_i \\ \mathbf{v}=\mathbf{v}_0}} \quad (37)$$

where $\mathbf{v}_0$ represents the vector of actual controller parameters, $\sigma_i$ means actual poles and $\Delta\sigma_i$ and $\Delta v_j$ are increments of poles and controller parameters, respectively. In case of a $p$-multiple pole, the following term is inserted in (36) and (37) instead of $m(s)$

$$\frac{\mathrm{d}^p}{\mathrm{d}s^p} m(s) \quad (38)$$

However, (38) can be used only if the pole including all multiplicities is moved. If, on the other hand, the intention is to shift a part of poles within the multiplicity to the one location and the rest of the multiplicity to another (or other) location(s), it is better to consider a multiple pole as a "nest" of close single poles.

Then a matrix

$$\mathbf{S} = \left[ \frac{\Delta\sigma_i}{\Delta v_j} \right] \in \mathbb{R} \ n_{sp} x l_{den} \quad (39)$$

is called the sensitivity matrix satisfying

$$\Delta\mathbf{v} = \mathbf{S}^+ \Delta\boldsymbol{\sigma} \quad (40)$$

where $\Delta\boldsymbol{\sigma} = \left[ \Delta\sigma_1, \Delta\sigma_2, ..., \Delta\sigma_{n_{sp}} \right]^T$ and $\mathbf{S}^+$ means the pseudoinverse.

It holds that

$$\frac{\Delta \mathrm{Re}\{\sigma_i\}}{\Delta v_j} = \mathrm{Re}\left\{ \frac{\Delta\sigma_i}{\Delta v_j} \right\} \quad (41)$$

thus, if poles are shifted in a real axis only, it can be calculated

$$\Delta\mathbf{v} = \mathrm{Re}\{\mathbf{S}\}^+ \mathrm{Re}\{\Delta\boldsymbol{\sigma}\} \quad (42)$$

Otherwise, the following approximation ought to be used

$$\Delta\mathbf{v} \approx \mathrm{Re}\{\mathbf{S}^+ \Delta\boldsymbol{\sigma}\} \quad (43)$$

The continuous shifting starts with $n_{sp} = \bar{n}_{den}$. Then, one can take the number of $\bar{n}_{den}$ rightmost poles and move them to the prescribed ones. The rightmost closed-loop pole moves to the rightmost prescribed pole etc. Alternatively, the same number of dominant poles (or conjugated pairs) can be considered; however, the nearest poles can be shifted to the prescribed ones. If two or more prescribed poles own the same dominant pole, it is assigned to the rightmost prescribed pole and removed from the list of moved poles. The number $n_{sp} \in \{\bar{n}_{den}, l_{den}\}$ is incremented whenever the approaching starts to fail for any pole. If $n_{sp} > \bar{n}_{den}$, the rest of dominant poles is pushed to the left. More precisely, shifting to the prescribed poles is described by the following formula

$$\Delta\sigma = \frac{\sigma_p - \sigma_s}{|\sigma_p - \sigma_s|}\delta \tag{44}$$

and pushing to the left agrees with

$$\Delta\sigma = -\delta \tag{45}$$

where $\delta$ is a discretization step in the space of poles, e.g. $\delta = 0.001$, $\sigma_p$ is a prescribed pole and $\sigma_s$ means a pole moved to the prescribed one.

If $n_{sp} = l_{den}$ and all prescribed poles become the rightmost (dominant) ones, PPSA is finished. Otherwise, do the last step of PPSA introduced in the following subsection.

## 5. Minimization of a cost function via SOMA

This step is implemented whenever the exact pole assignment even via shifting fails. In the first part of this subsection we arrange the cost function to be minimized. Then, SOMA algorithm (Zelinka, 2004) belonging to the wide family of evolution algorithms is introduced and briefly described. Again, the procedure is given for the pole-optimization; the zero-optimization dealing with the closed-loop numerator is done analogously.

### 5.1 Cost function

The goal now is to rearrange feedback poles (zeros) so that they are "sufficiently close" to the prescribed ones and, concurrently, they are "as the most dominant as possible". This requirement can be satisfied by the minimizing of the following cost function

$$F(\mathbf{v}) = d_\sigma(\mathbf{v}) + \lambda d_R(\mathbf{v}) = \sum_{i=1}^{\bar{n}_{den}} |\sigma_{s,i} - \sigma_{p,i}| + \lambda \operatorname{Re}\{\sigma_{d,i} - \sigma_{p,i}\} \tag{46}$$

where $d_\sigma(\mathbf{v})$ is the distance of prescribed poles $\sigma_{p,i}$ from the nearest ones $\sigma_{s,i}$, $d_R(\mathbf{v})$ expresses the sum of distances of dominant poles from the prescribed ones and $\lambda > 0$ represents a real weighting parameter. The higher $\lambda$ is, the pole dominancy of is more important in $F(\mathbf{v})$. Recall that (when the dominant poles were moved)

$$\sigma_{s,1} \geq \sigma_{s,2} \geq \ldots \geq \sigma_{s,\bar{n}_{den}}, \sigma_{p,1} \geq \sigma_{p,2} \geq \ldots \geq \sigma_{p,\bar{n}_{den}}, \sigma_{d,1} \geq \sigma_{d,2} \geq \ldots \geq \sigma_{d,\bar{n}_{den}} \tag{47}$$

Alternatively, one can include both, the zeros and poles, in (46), not separately.

Poles can be found e.g. by the quasipolynomial mapping root finder (QPMR) implemented in Matlab, see (Vyhlídal & Zítek, 2003).

Hence, the aim is to solve the problem

$$\mathbf{v}_{opt} = \arg\min F(\mathbf{v}) \qquad (48)$$

We use SOMA algorithm based on genetic operations with a population of found solutions and moving of population specimens to each other. A brief description of the algorithm follows.

## 5.2 SOMA

SOMA is ranked among evolution algorithms, more precisely genetic algorithms, dealing with populations similarly as differential evolution does. The algorithm is based on vector operations over the space of feasible solutions (parameters) in which the population is defined. Population specimens cooperate when searching the best solution (the minimum of the cost function) and, simultaneously, each of them tries to be a leader. They move to each other and the searching is finished when all specimens are localized on a small area.

In SOMA, every single generation, in which a new population is generated, is called a migration round. The notion of specific control and termination parameters, which have to be set before the algorithm starts, will be explained in every step of a migration round below.

First, population $P = \{\mathbf{v}_1, \mathbf{v}_2, ..., \mathbf{v}_{PopSize}\}$ must be generated based on a prototypal specimen. For PPSA, this specimen is a vector of controller free parameters, $\mathbf{v}$, of dimension $D = l_{den}$. The prototypal specimen equals the best solution from Subsection 4.5. One can choose an initial radius (*Rad*) of the population in which other specimens are generated. The size of population (*PopSize*), i.e. the number of specimens in the population, is chosen by the user. Each specimen is then evaluated by the cost function (46).

The simplest strategy called "All to One" implemented here then selects the best specimen - leader, i.e. that with the minimal value of the cost function

$$\mathbf{v}_L^{mr} = \arg\min_i F\left(\mathbf{v}_i^{mr}\right) \qquad (49)$$

where $L$ denotes the leader, $i$ is $i$-th of specimen in the population and $mr$ means the current migration round. Then all other specimen are moved towards the leader during the migration round. The moving is given by three control parameters: *PathLength, Step, PRT*. *PathLength* should be within the interval [1.1,5] and it expresses the length of the path when approaching the leader. *PathLength* = 1 means that the specimen stops its moving exactly at the position of the leader. *Step* represents the sampling of the path and ought to be valued $[0.11, PathLength]$. E.g. a pair *PathLength* = 1 and *Step* = 0.2 agrees with that the specimen makes 5 steps until it reaches the leader. $PRT \in [0,1]$ enables to calculate the perturbation vector *PRTVector* which indicates whether the active specimen moves to the leader directly or not. *PRTVector* is defined as

$$PRTVector = \left[p_1, p_2, ..., p_{l_{den}}\right]^T \in \{0,1\}^{l_{den}}$$
$$p_i = 1 \quad \text{if } rnd_i < PRT \qquad (50)$$
$$p_i = 0 \quad \text{else}$$

where $rnd_i \in [0,1]$ is a randomly generated number for each dimension of a specimen. Although authors of SOMA suggest to calculate $PRTVector$ only once in migration round for every specimen, we try to do this in every step of the moving to the leader. Hence, the path is given by

$$\mathbf{v}_{i,k+1}^{mr} = \mathbf{v}_{i,0}^{mr} + (i-1)Step\left(\mathbf{v}_L^{mr} - \mathbf{v}_{i,0}^{mr}\right) + \mathrm{diag}\left(PRTVector\right)Step\left(\mathbf{v}_L^{mr} - \mathbf{v}_{i,0}^{mr}\right)$$
$$i = 1,2,...PopSize; k = 0,1,...round\left(PathLength\,/\,Step\right) - 1$$
(51)

where $\mathrm{diag}\left(PRTVector\right)$ means the diagonal square matrix with elements of $PRTVector$ on the main diagonal and $k$ is $k$-th step in the path.

If $PRTVector = [1,1,...1]^T$, the active specimen goes to the leader directly without "zig-zag" moves.

For every specimen of the population in a migration round, the cost function (i.e. value of the specimen) is calculated in every single step during the moving towards the leader. If the current position is better then the actual best, it becomes the best now. Hence, the new position of an active specimen for the next migration round is given by the best position of the specimen from all steps of moving towards the leader within the current migration round, i.e.

$$\mathbf{v}_i^{mr+1} = \arg\min_k F\left(\mathbf{v}_{i,k}^{mr}\right)$$
(52)

These specimens then generate the new population.

The number of migration round are given by user at the beginning of SOMA by parameter *Migration*, or the algorithm is terminated if

$$\max_i F\left(\mathbf{v}_i\right) - \min_i F\left(\mathbf{v}_i\right) < MinDiv$$
(53)

where $MinDiv$ is the selected minimal diversity.

The final value $\mathbf{v}_{opt}$ is equal to $\mathbf{v}_L$ from the last migration round. We implemented the whole PPSA with SOMA in two Matlab m-files.

## 6. Illustrative example

In this closing session, we demonstrate the utilization of the PPSA and the methodology described above in Matlab on an attractive example.

Consider an unstable system describing roller skater on a swaying bow (Zítek et al., 2008) given by the transfer function

$$G(s) = \frac{Y(s)}{U(s)} = \frac{b\exp\left(-(\tau + \vartheta)s\right)}{s^2\left(s^2 - a\exp(-\vartheta s)\right)}$$
(54)

see Fig. 8, where $y(t)$ is the skater's deviation from the desired position, $u(t)$ expresses the slope angle of a bow caused by force $P$, delays $\tau, \vartheta$ mean the skater's and servo latencies and $b, a$ are real parameters. Skater controls the servo driving by remote signals into servo electronics.

Let $b = 0.2$, $a = 1$, $\tau = 0.3$ s, $\vartheta = 0.1$ s, as in the literature, and design the controller structure according to the approach described in Section 3. Consider the reference and load disturbance in the form of a step-wise function.

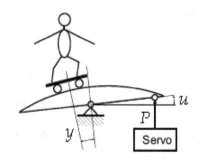

Fig. 8. The roller skater on a swaying bow

Hence, coprime factorization over $\mathbf{R}_{MS}$ can be done e.g. as

$$G(s) = \frac{B(s)}{A(s)} = \frac{\dfrac{b\exp\left(-(\tau+\vartheta)s\right)}{(s+m_0)^4}}{\dfrac{s^2\left(s^2 - a\exp(-\vartheta s)\right)}{(s+m_0)^4}}, W(s) = \frac{H_W(s)}{F_W(s)} = \frac{\dfrac{k_W}{s+m}}{\dfrac{s}{s+m}}, D(s) = \frac{H_D(s)}{F_D(s)} = \frac{\dfrac{k_D}{s+m}}{\dfrac{s}{s+m}} \quad (55)$$

where $m_0 > 0$, $k_W$, $k_D \in \mathbb{R}$. Stabilization via the Bézout identity (15) results e.g. in the following particular solution

$$Q_0(s) = \frac{\left(q_3 s^3 + q_2 s^2 + q_1 s + q_0\right)(s+m_0)^4}{s^2\left(s^2 - a\exp(-\vartheta s)\right)\left(s^3 + p_2 s^2 + p_1 s + p_0\right) + b\exp\left(-(\tau+\vartheta)s\right)\left(q_3 s^3 + q_2 s^2 + q_1 s + q_0\right)}$$

$$P_0(s) = \frac{\left(s^3 + p_2 s^2 + p_1 s + p_0\right)(s+m_0)^4}{s^2\left(s^2 - a\exp(-\vartheta s)\right)\left(s^3 + p_2 s^2 + p_1 s + p_0\right) + b\exp\left(-(\tau+\vartheta)s\right)\left(q_3 s^3 + q_2 s^2 + q_1 s + q_0\right)}$$

$$(56)$$

using the generalized Euclidean algorithm, see (Pekař & Prokop, 2009), where $p_2$, $p_1$, $p_0$, $q_3$, $q_2$, $q_1$, $q_0 \in \circ$ are free parameters. In order to provide reference tracking and load disturbance rejection, use parameterization (16) while both, $F_W(s)$ and $F_D(s)$, divide $P(s)$; in other words, the numerator of $P(s)$ must satisfy $P(0) = 0$. If we take

$$T = \frac{t_0(s+m_0)^4}{s^2\left(s^2 - a\exp(-\vartheta s)\right)\left(s^3 + p_2 s^2 + p_1 s + p_0\right) + b\exp\left(-(\tau+\vartheta)s\right)\left(q_3 s^3 + q_2 s^2 + q_1 s + q_0\right)} \quad (57)$$

$P(s)$ is obtained in a quite simple form with a real parameter $t_0$ which must be set as

$$t_0 = \frac{-p_0 m_0^4}{b} \quad (58)$$

Finally, the controller numerator and denominator in $\mathbf{R}_{MS}$, respectively, have forms

$$Q(s) = \frac{b\left(q_3 s^3 + q_2 s^2 + q_1 s + q_0\right)(s + m_0)^4 + p_0 m_0^4 s^2 \left(s^2 - a\exp(-\vartheta s)\right)}{b\left[s^2\left(s^2 - a\exp(-\vartheta s)\right)\left(s^3 + p_2 s^2 + p_1 s + p_0\right) + b\exp\left(-(\tau + \vartheta)s\right)\left(q_3 s^3 + q_2 s^2 + q_1 s + q_0\right)\right]}$$ (59)

$$P(s) = \frac{\left(s^3 + p_2 s^2 + p_1 s + p_0\right)(s + m_0)^4 - p_0 m_0^4 \exp\left(-(\tau + \vartheta)s\right)}{s^2\left(s^2 - a\exp(-\vartheta s)\right)\left(s^3 + p_2 s^2 + p_1 s + p_0\right) + b\exp\left(-(\tau + \vartheta)s\right)\left(q_3 s^3 + q_2 s^2 + q_1 s + q_0\right)}$$

Hence, the controller has the transfer function

$$G_R(s) = \frac{b\left(q_3 s^3 + q_2 s^2 + q_1 s + q_0\right)(s + m_0)^4 + p_0 m_0^4 s^2 \left(s^2 - a\exp(-\vartheta s)\right)}{b\left[\left(s^3 + p_2 s^2 + p_1 s + p_0\right)(s + m_0)^4 - p_0 m_0^4 \exp\left(-(\tau + \vartheta)s\right)\right]}$$ (60)

and the reference-to-output function reads

$$G_{WY}(s)$$

$$= \frac{b\left[b\left(q_3 s^3 + q_2 s^2 + q_1 s + q_0\right)(s + m_0)^4 + p_0 m_0^4 s^2 \left(s^2 - a\exp(-\vartheta s)\right)\right]\exp\left(-(\tau + \vartheta)s\right)}{(s + m_0)^4\left[s^2\left(s^2 - a\exp(-\vartheta s)\right)\left(s^3 + p_2 s^2 + p_1 s + p_0\right) + b\exp\left(-(\tau + \vartheta)s\right)\left(q_3 s^3 + q_2 s^2 + q_1 s + q_0\right)\right]}$$ (61)

which gives rise to the characteristic quasipolynomial

$$m(s)$$

$$= (s + m_0)^4\left[s^2\left(s^2 - a\exp(-\vartheta s)\right)\left(s^3 + p_2 s^2 + p_1 s + p_0\right) + b\exp\left(-(\tau + \vartheta)s\right)\left(q_3 s^3 + q_2 s^2 + q_1 s + q_0\right)\right]$$ (62)

Obviously, the numerator of $G_{WY}(s)$ does not have any free parameter not included in the denominator, i.e. $l_{num} = 0$. Moreover, the factor $(s + m_0)^4$ has a quadruple real pole; to cancel it, it must hold that $m_0 \gg -\mathrm{Re}\{s_1\} = -\alpha$. Hence $l_{den} = 7$. Now, there are two possibilities – either set zero exactly to obtain constrained controller parameter (then $l_{den} = 6$) or to deal with the numerator and denominator of (61) together in (46) – we decided to utilize the former one. Generally, one can obtain e.g.

$$p_0 = -\frac{b(z_1 + m_0)^4\left(q_3 z_1^3 + q_2 z_1^2 + q_1 z_1 + q_0\right)}{m_0^4 z_1^2\left(z_1^2 - a\exp(-\vartheta z_1)\right)}$$ (63)

from (61).

Choose $\Delta h_{WY,m,\max} = 0.5$, $\xi_\alpha = 0.5$ and $t_{\max} = 10$ s. From Fig. 3 we have $\xi_z = 0.9$, $t_{\max,norm} \approx 2$ which gives $\omega = 0.2, z_1 = -0.18, \alpha = -0.1$. Then take e.g. $m_0 = 5$. Inserting plant parameters in (63) yields

$$p_0 = 5.4078\left(q_0 - 0.18 q_1 + 0.0324 q_2 - 0.005832 q_3\right)$$ (64)

The concrete quasipolynomial which roots are being set, thus, reads

$$m_1(s) = s^2\left(s^2 - \exp(-0.1s)\right)\left(s^3 + p_2 s^2 + p_1 s + 5.4078\left(q_0 - 0.18 q_1 + 0.0324 q_2 - 0.005832 q_3\right)\right)$$
$$+ 0.2\exp(-0.4s)\left(q_3 s^3 + q_2 s^2 + q_1 s + q_0\right)$$ (65)

Initial direct pole placement results in controller parameters as

$$q_3 = 1.0051, q_2 = 0.9506, q_1 = 1.2582, q_0 = 0.2127, p_2 = 1.1179, p_1 = 0.4418, p_0 = 0.0603 \quad (66)$$

and poles locations in the vicinity of the origin are displayed in Fig. 9.

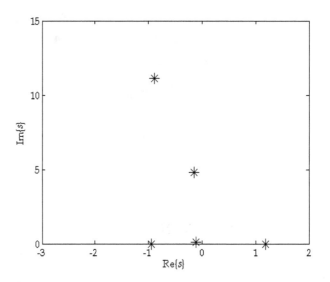

Fig. 9. Initial poles locations

The process of continuous roots shifting is described by the evolution of controller parameters, the spectral abscissa (i.e. the real part of the rightmost pole $\sigma_{d,1}$) and the distance of the dominant pole from the prescribed one $\left|\sigma_{d,1} - \sigma_{p,1}\right|$, as can be seen in Fig. 10 – Fig. 12, respectively. Note that $p_0$ is related to shifted parameters according to (64).

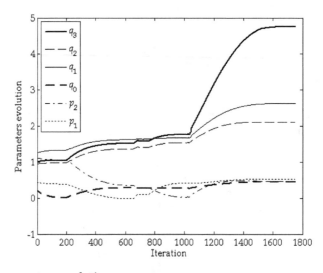

Fig. 10. Shifted parameters evolution

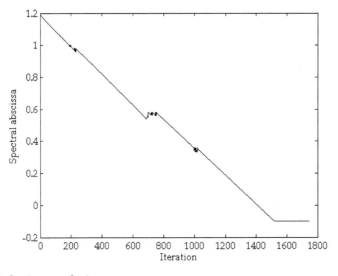

Fig. 11. Spectral abscissa evolution

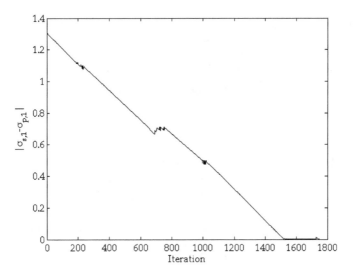

Fig. 12. Distance of the rightmost pole from the prescribed one

When shifting, it is suggested to continue in doing this even if the desired poles locations are reached since one can obtain a better poles distribution – i.e. non-dominant poles are placed more left in the complex space. Moreover, one can decrease the number of shifted poles during the algorithm whenever the real part of a shifted pole becomes "too different" from a group of currently moved poles.

The final controller parameters from the continuous shifting are

$$q_3 = 4.7587, q_2 = 2.1164, q_1 = 2.6252, q_0 = 0.4482, p_2 = 0.4636, p_1 = 0.529, p_0 = 4.6164 \quad (67)$$

and the poles location is pictured in Fig. 13.

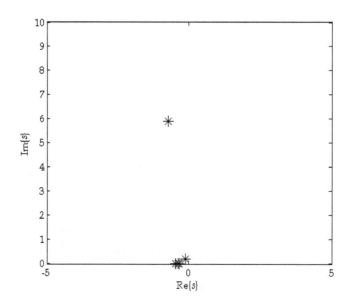

Fig. 13. Final poles locations

As can be seen, the desired prescribed pole is reached and it is also the dominant one. Thus, optimization can be omitted. However, try to perform SOMA to find the minimal cost function (16) with this setting: *Rad* = 2, *PopSize* = 10, *D* = 6, *PathLength* = 3, *Step* = 0.21, *PRT* = 0.6, *Migration* = 10, *MinDiv* = 10⁻⁶, . Yet, the minimum of the cost function remains in the best solution from continuous shifting, i.e. according to (67), with the value of the cost function as $F(\mathbf{v}) = 2.93 \cdot 10^{-4}$ .

## 7. Conclusion

This chapter has introduced a novel controller design approach for SISO LTI-TDS based on algebraic approach followed by pole-placement-like controller tuning and an optimization procedure. The methodology has been implemented in Matlab-Simulink environment to verify the results.

The initial controller structure design has been made over the ring of stable and proper meromorphic functions, $\mathbf{R}_{MS}$, which offers to satisfy properness of the controller, reference tracking and rejection of the load disturbance (of a nominal model). The obtained controller has owned some free (unset) parameters which must have been set properly.

In the crucial part of the work, we have chosen a simple finite-dimension model, calculated its step-response maximum overshoots and times to the overshoots. Then, using a static pole placement followed by continuous pole shifting dominant poles have been attempted to be placed to the desired prescribed positions.

Finally, optimization of distances of dominant (the rightmost) poles from the prescribed ones has been utilized via SOMA algorithm. The whole methodology has been tested on an attractive example of a skater on a swaying bow described by an unstable LTI TDS model.

The procedure is similar to the algorithm introduced in (Michiels et al., 2010); however, there are some substantial differences between them. Firstly, the presented approach is made in input-output space of meromorphic Laplace transfer functions, whereas the one in (Michiels et al., 2010) deals purely with state space. Second, in the cited literature, a number poles less then a number of free controller parameters is set exactly and the rest of the spectrum is pushed to the left as much as possible. If it is possible it is necessary to choose other prescribed poles. We initially place the poles exactly; however, they can leave their positions during the shifting. Anyway, our algorithm does not require reset of selection assigned poles. Moreover, we suggest unambiguously how the prescribed poles (and zeros) positions are to be chosen – based on model overshoots. Last but not least, in (Michiels et al., 2010), the gradient sampling algorithm (Burke et al., 2005) on the spectral abscissa was used while SOMA together with more complex cost function is considered in this chapter.

The presented approach is limited to retarded SISO LTI-TDS without distributed delays only. Its extension to neutral systems requires some additional conditions on stability and existence of a stabilizing controller. Systems with distributed delays can be served in similar way as it is done here, yet with the characteristic meromorphic function instead of quasipolynomial. Multivariable systems would require deeper theoretic analysis of the controller structure design. The methodology is also time-comsupting and thus useless for online controller design (e.g. for selftuners).

In the future research, one can solve the problems specified above, choose other reference-to-output models and control system structures. There is a space to improve and modify the optimization algorithm.


## 8. Acknowledgment

The authors kindly appreciate the financial support which was provided by the Ministry of Education, Youth and Sports of the Czech Republic, in the grant No. MSM 708 835 2102 and by the European Regional Development Fund under the project CEBIA-Tech No. CZ.1.05/2.1.00/03.0089.


## 9. References

Ben-Israel, A. & Greville, T.N.E (1966). *Generalized Inverses: Theory and Applications*. Wiley, New York

Burke, J.; Lewis, M & Overton, M. (2005). A Robust Gradient Sampling Algorithm for Nonsmooth Nonconvex Optimization. *SIAM Journal of Optimization*, Vol. 15, Issue 3, pp. 751-779, ISSN 1052-6234

El'sgol'ts, L.E. & Norkin, S.B. (1973). *Introduction to the Theory and Application of Differential Equations with Deviated Arguments*. Academic Press, New York

Górecki, H.; Fuksa, S.; Grabowski, P. & Korytowski, A. (1989). *Analysis and Synthesis of Time Delay Systems*, John Wiley & Sons, ISBN 978-047-1276-22-7, New York

Hale, J.K., & Verduyn Lunel, S.M. (1993). Introduction to Functional Differential Equations. *Applied Mathematical Sciences*, Vol. 99, Springer-Verlag, ISBN 978-038-7940-76-2, New York

Kolmanovskii, V.B. & Myshkis, A. (1999). *Introduction to the Theory and Applications of Functional Differential Equations*, Cluwer Academy, ISBN 978-0792355045, Dordrecht, Netherlands

Kučera, V. (1993). Diophantine equations in control - a survey. Automatica, Vol. 29, No. 6, pp. 1361-1375, ISSN 0005-1098

Marshall, J.E.; Górecki, H.; Korytowski, A. & Walton, K. (1992). *Time Delay Systems, Stability and Performance Criteria with Applications*. Ellis Horwood, ISBN 0-13-465923-6, New York

Michiels, W.; Engelborghs, K; Vansevenant, P. & Roose, D. (2002). Continuous Pole Placement for Delay Equations. *Automatica*, Vol. 38, No. 6, pp. 747-761, ISSN 0005-1098

Michiels, W.; Vyhlídal, T. & Zítek, P. (2010). Control Design for Time-Delay Systems Based on Quasi-Direct Pole Placement. *Journal of Process Control*, Vol. 20, No. 3, pp. 337-343, 2010.

Loiseau, J.-J. (2000). Algebraic Tools for the Control and Stabilization of Time-Delay Systems. *Annual Reviews in Control*, Vol. 24, pp. 135-149, ISSN 1367-5788

Loiseau, J.-J. (2002). Neutral-Type Time-Delay Systems that are not Formally Stable are not BIBO Stabilizable. *IMA Journal of Mathematical Control and Information*, Vol. 19, No. 1-2, pp. 217-227, ISSN 0265-0754

Michiels, W. & Niculescu, S. (2008). Stability and Stabilization of Time Delay Systems: An eigenvalue based approach. *Advances in Design and Control*, SIAM, ISBN 978-089-8716-32-0, Philadelphia

Pekař, L. & Prokop, R. (2009). Some observations about the RMS ring for delayed systems, Proceedings of the 17th International Conference on Process Control '09, pp. 28-36, ISBN 978-80-227-3081-5, Štrbské Pleso, Slovakia, June 9-12, 2009.

Pekař, L.; Prokop, R. & Dostálek, P. (2009). Circuit Heating Plant Model with Internal Delays. *WSEAS Transaction on Systems*, Vol. 8, Issue 9, pp. 1093-1104, ISSN 1109-2777

Pekař, L. & Prokop, R. (2010). Control design for stable systems with both input-output and internal delays by algebraic means, *Proceedings of the 29th IASTED International Conference Modelling, Identification and Control (MIC 2010)*, pp. 400-407, ISSN 1025-8973, Innsbruck, Austria, February 15-17, 2010

Richard, J.P. (2003). Time-delay Systems: an Overview of Some Recent Advances and Open Problems. *Automatica*, Vol. 39, Issue 10, pp. 1667-1694, ISSN 0005-1098

Vyhlídal, T. (2003). *Analysis and Synthesis of Time Delay System Spectrum*, Ph.D. Thesis, Faculty of Mechanical Engineering, Czech Technical University in Prague, Prague

Vyhlídal, T. & Zítek, P. (2003). Quasipolynomial mapping based rootfinder for analysis of time delay systems, *Proceedings IFAC Workshop on Time-Delay systems, TDS'03*, pp. 227-232, Rocquencourt, France 2003.

Zelinka, I. (2004). SOMA-Self Organizing Migrating Algorithm. In: *New Optimization Techniques in Engineering*, G. Onwubolu, B.V.Babu (Eds.), 51 p., Springer-Verlag, ISBN 3-540-20167X, Berlin

Zítek, P. & Kučera, V. (2003). Algebraic Design of Anisochronic Controllers for Time Delay Systems. *International Journal of Control*, Vol. 76, Issue 16, pp. 905-921, ISSN 0020-7179

Zítek, P. & Hlava, J. (2001). Anisochronic Internal Model Control of Time-Delay Systems. *Control Engineering Practice*, Vol. 9, No. 5, pp. 501-516, ISSN 0967-0661

Zítek, P.; Kučera, V. & Vyhlídal, T. (2008). Meromorphic Observer-based Pole Assignment in Time Delay Systems. *Kybernetika*, Vol. 44, No. 5, pp. 633-648, ISSN 0023-5954

# Permissions

All chapters in this book were first published by InTech Open; hereby published with permission under the Creative Commons Attribution License or equivalent. Every chapter published in this book has been scrutinized by our experts. Their significance has been extensively debated. The topics covered herein carry significant findings which will fuel the growth of the discipline. They may even be implemented as practical applications or may be referred to as a beginning point for another development.

The contributors of this book come from diverse backgrounds, making this book a truly international effort. This book will bring forth new frontiers with its revolutionizing research information and detailed analysis of the nascent developments around the world.

We would like to thank all the contributing authors for lending their expertise to make the book truly unique. They have played a crucial role in the development of this book. Without their invaluable contributions this book wouldn't have been possible. They have made vital efforts to compile up to date information on the varied aspects of this subject to make this book a valuable addition to the collection of many professionals and students.

This book was conceptualized with the vision of imparting up-to-date information and advanced data in this field. To ensure the same, a matchless editorial board was set up. Every individual on the board went through rigorous rounds of assessment to prove their worth. After which they invested a large part of their time researching and compiling the most relevant data for our readers.

The editorial board has been involved in producing this book since its inception. They have spent rigorous hours researching and exploring the diverse topics which have resulted in the successful publishing of this book. They have passed on their knowledge of decades through this book. To expedite this challenging task, the publisher supported the team at every step. A small team of assistant editors was also appointed to further simplify the editing procedure and attain best results for the readers.

Apart from the editorial board, the designing team has also invested a significant amount of their time in understanding the subject and creating the most relevant covers. They scrutinized every image to scout for the most suitable representation of the subject and create an appropriate cover for the book.

The publishing team has been an ardent support to the editorial, designing and production team. Their endless efforts to recruit the best for this project, has resulted in the accomplishment of this book. They are a veteran in the field of academics and their pool of knowledge is as vast as their experience in printing. Their expertise and guidance has proved useful at every step. Their uncompromising quality standards have made this book an exceptional effort. Their encouragement from time to time has been an inspiration for everyone.

The publisher and the editorial board hope that this book will prove to be a valuable piece of knowledge for researchers, students, practitioners and scholars across the globe.

# List of Contributors

**Valeria Boscaino and Giuseppe Capponi**
University of Palermo, Italy

**Cornelis Jan Kikkert**
James Cook University, Queensland, Australia

**Ricardo Vargas, M.A Arjona and Manuel Carrillo**
Instituto Tecnológico de la Laguna, División de Estudios de Posgrado e Investigación, México

**Peter Drgoňa, Michal Frivaldský and Anna Simonová**
University of Žilina, Faculty of Electrical Engineering, Žilina, Slovakia

**Lutfi Al-Sharif, Mohammad Kilani, Sinan Taifour, Abdullah Jamal Issa, Eyas Al-Qaisi, Fadi Awni Eleiwi and Omar Nabil Kamal**
Mechatronics Engineering Department, University of Jordan, Jordan

**Iman Mohammad Hoseiny Naveh and Javad Sadeh**
Islamic Azad University, Gonabad Branch. Islamic Republic of Iran

**Anibal Azevedo**
State University of São Paulo, Brazil

**Libor Pekař and Roman Prokop**
Tomas Bata University in Zlín, Czech Republic

# Index

Printed in the USA
CPSIA information can be obtained
at www.ICGtesting.com
JSHW051349091023
49903JS00006B/85

9 781639 876945